CREATION

The Story of the Origin and Evolution of the Universe

Barry Parker

Drawings by
Lori Scoffield

PLENUM PRESS • NEW YORK AND LONDON

Library of Congress Cataloging in Publication Data

Parker, Barry R.
Creation: the story of the origin and evolution of the universe / Barry R. Parker; drawings by Lori Scoffield.
p. cm.
Bibliography: p.
Includes index.
ISBN 0-306-42952-7
1. Cosmology. I. Scoffield, Lori. II. Title.
QB981.P29 1988 88-17893
523.1—dc19 CIP

Plenum Press is a Division of
Plenum Publishing Corporation
233 Spring Street, New York, N.Y. 10013

Printed in the United States of America

CREATION

The Story of the Origin and Evolution of the Universe

OTHER BOOKS BY BARRY PARKER

EINSTEIN'S DREAM
The Search for a Unified Theory of the Universe

SEARCH FOR A SUPERTHEORY
From Atoms to Superstrings

Preface

I remember once watching a presentation of the creation of the universe in a planetarium. It was a fascinating experience: lights flashing, particles appearing to rush by as an explosive roar echoed throughout the planetarium. Then suddenly . . . blackness. And after a few seconds . . . tiny lights—stars blinking into existence. I tried to imagine myself actually going back to this event. Was this really what it was like? It was an interesting facsimile, but far from what the real thing would have been like. The creation of the universe is an event that is impossible to imagine accurately. Fortunately, this has not discouraged people from wondering what it was like.

In *Creation* I have attempted to take you back to the beginning—the big bang explosion—so that you can watch the universe grow and evolve. Starting with the first fraction of a second, I trace the universe from its initial dramatic expansion through to the formation of the first nuclei and atoms. From here I go to the formation of galaxies and the curious distribution they have taken in space. Finally I talk about the formation of elements in stars, and the first life on the planets around them.

It's a fascinating story, but the story I tell is not just about the universe. It's also about the scientists who made the discoveries—Einstein, Hubble, Gamow, Guth, and many others. Hundreds of scientists made contributions, and there is no doubt an interesting story associated with each of them. And I must apologize for all of those whom I was not able to include.

Some of the material of this book is still speculative and changes will no doubt occur in coming years. This should not surprise you, though. We are dealing with the frontiers of science—and frontiers are never static. It sometimes takes years to thoroughly test a new idea. Even the existence of atoms was controversial for decades. Yet now, despite never seeing one, we feel confident that they do indeed exist.

It is not possible to write such a story without occasionally using scientific terms. I have tried to explain each of them as they arise, but for the benefit of those new to science I have included a glossary.

Very large and very small numbers are also a problem. It is difficult to talk about the objects in astronomy without using them. To get around writing them out explicitly I have used what is called scientific notation. In this notation the number 100,000 is written as 10^5 (i.e., the index gives the number of zeros after the one). For very small numbers such as 1/10,000 I write 10^{-4}. Temperature scales may also bother you. Most of you, no doubt, feel most comfortable with the Fahrenheit scale. Astronomers, however, prefer to use the Kelvin scale (abbreviated as K). On this scale the lowest temperature in the universe is 0 K. On the Fahrenheit scale this is −459°F. The boiling point of water on this scale is 373 K.

I am particularly grateful to the scientists who assisted me. Interviews were conducted either in person, over the telephone, or via letter with many of the people mentioned in the book. I also thank many of them for photos and reprints. They are: Ralph Alpher, Michael Turner, David Schramm, Alan Guth, Jim Peebles, Paul Steinhardt, Andreas Albrecht, John Huchra, Gary Steigman, Leonard Susskind, Frank Wilczek, William Fowler, Robert Wagoner, Arno Penzias, Ed Tryon, Jim Hartle, Alex Vilenkin, A. Linde, Jeremiah Ostriker, Edward Kolb, and Heinz Pagels.

The sketches and some of the line drawings were done by

Lori Scoffield. The remainder of the line drawings were done by students at ISU Vo-Tech. I thank all of them for an excellent job. I also thank Linda Greenspan Regan, Victoria Cherney, and the staff of Plenum for their assistance in bringing the book into its final form. And finally I would like to thank my wife for her support during the writing of the book.

Barry Parker

Contents

CHAPTER 1

Introduction

Through a telescope the night sky comes alive. Stars—millions upon millions of them—dot the darkness, twinkling like tiny beacons in a vast ocean. Doublets, triplets, and even assemblages of thousands dance in the field of view as you scan the darkness. Some are blue, some red, others yellow like our sun. Some are so large that if they took the place of our sun, they would engulf the Earth. And others are hardly larger than Jupiter. Around many of these stars are systems of planets, perhaps similar to our own solar system. And in the vast regions between the stars lie colorful glowing lagoons of gas coupled with huge dark clouds of gas and dust. This is the birthplace of stars, and even now stars are forming there.

Some of these stars are expanding and contracting rhythmically, varying in brightness over days, months, and even years. Others—pulsars—are like tiny lighthouses, blinking on, off, on, off in seconds. And occasionally, massive stars explode in the greatest fireworks display imaginable—the supernova. The remnants of such an explosion then go into the making of new stars—second-generation stars, then third generation, and so on.

If you could venture into space in a spaceship you might find an even more exotic object—a black hole. Only a few miles across, it emits no light, but if you were close enough you would see it as a black circle outlined on the star clouds behind it. If you were to approach too close you would be pulled into it. And once you passed through its surface there would be no escape.

Stars and gaseous nebula within the Milky Way (interior of the Rosette nebula). (Courtesy National Optical Astronomy Observatories.)

This is the island universe of stars, the galaxy that we live in—the Milky Way. It contains about 200 billion stars and is so large that even if you could travel at the speed of light, 100,000 years would be needed to traverse it. Shaped like a disk with spiral arms, it takes 200,000 years to spin once on its axis. Looking upward on a clear summer evening you can see it as a faint ribbon of light strung across the sky. We are, of course, seeing it from the inside—from a position about two-thirds the way out from the center.

If we look beyond our galaxy we see other galaxies: one in the constellation (group of stars) Andromeda, another in the nearby constellation Triangulum. Some of these galaxies are spirals, like ours. Others are elliptical and a few are completely

The Eta Carinae nebula. (Courtesy National Optical Astronomy Observatories.)

irregular in shape, like the Magellanic clouds. In all, there are hundreds of billions of galaxies, each containing hundreds of billions of stars. And beyond the farthest galaxies are the enigmatic quasars—objects that have puzzled astronomers for over twenty years. They are tiny, perhaps no bigger than our solar system, but they give out more energy than a thousand galaxies.

This is the universe—a structure so large that it is beyond human comprehension. Earth is but a tiny speck within it, lost in the glare of its star near the outskirts of one of the hundreds of billions of galaxies. Why, we wonder, are we here? Are we the only humans in this vastness? And perhaps the most important question of all: Where did the universe come from?

One possible answer is that the universe has always been here. But scientists do not buy this. Evidence, accumulated over

The Tarantula nebula in the Large Magellanic cloud. (Courtesy National Optical Astronomy Observatories.)

the last few decades, indicates that the universe came into being in an explosion of unimaginable intensity about 18 billion years ago. What was this explosion like? Certainly there is no way we could visualize it—that is beyond our powers of imagination. Indeed, only a few years ago most astronomers would have laughed if you told them we would soon be able to piece together the events that occurred in the first few seconds after this explosion. Yet it has happened. Scientists have been able to reach back almost to the first moment of the universe's existence. Many secrets remain locked away but an amazing story is unfolding—the story of creation itself.

As strange as it might seem, we can look almost all the way back to the "big bang" even now. How? All we have to do is go

An "island universe" of stars—a galaxy (the spiral galaxy in Triangulum). (Courtesy National Optical Astronomy Observatories.)

A cluster of galaxies in Hercules. (Courtesy National Optical Astronomy Observatories.)

to the telescope and look out into space. Because of the finite velocity of light (186,000 miles per second), the deeper you probe, the farther back in time you are looking. It takes a certain amount of time for the light from these objects to reach us: over four years from our nearest star (Alpha Centauri) and two million years from the bright galaxy in the constellation Andromeda. The stars in the night sky are therefore seen as they looked years ago, and the galaxies as they looked millions of years ago.

Using giant telescopes such as the two-hundred-inch one at Mount Palomar, we can probe back in time billions of years. If we probe back 18 billion years, we should, in theory, be able to see the first second of creation. Can we? No, but with the help of

Planet with a ring around it.

The four-meter telescope at Kitt Peak. (Courtesy National Optical Astronomy Observatories.)

Domes at Cerra Tololo Observatory (Chile). (Courtesy National Optical Astronomy Observatories.)

new techniques and larger, more powerful telescopes one day we may come very close.

So far I have been talking about looking back in time directly, but we can also look back another way. We can reconstruct the events that took place in the early universe through the use of mathematical formulae. In other words, we can use particle theories and Einstein's general theory of relativity.

How far back can we look? Can we actually "see" the creation of the universe in this way. We can come close but scientists are still having difficulties with creation itself. To see what the problem is, imagine that we are going back to the first second. Because the universe is expanding now, if we travel back in time it will contract. And as it contracts it will get smaller, hotter and denser. Eventually it will be so small, so dense, and so hot that Einstein's theory of relativity will no longer be valid, and we will have no theory to describe it. The particles that we know today will no longer exist—the universe will be a tiny, unbelievably hot primordial soup.

When does this breakdown occur? According to scientists it occurs at 10^{-43} second after the big bang. This is 1/1000000000000000000000000000000000000000000000th of a second. Needless to say, it is an incredibly short period of time. To most it might seem that we could neglect anything that happened before it. Unfortunately, a very critical event had happened—creation itself. And without a theory to explain this event we can only guess what happened. In particular, we cannot answer the question: What was the universe like at the very beginning? But scientists are human too . . . and they like to speculate. And they have speculated. John Wheeler of the University of Texas believes that it would have been like a bubbling froth, a foam with space and time in a tangled, disconnected array.

To further complicate things we have to think of this foam as inside a space smaller than an atom. With time severely distorted it may make little sense to ask what happened before this, for time, as we know it, did not exist.

How do we contemplate such a situation: the entire observable universe squeezed into a volume smaller than an atom, with no space around it? The big bang that created the universe created everything, including space. The only reasonable answer to this question is: we do not. Indeed, we cannot even make calculations describing it.

From this tiny nucleus the universe was born in an instant of unimaginable chaos. All the particles of nature—the electrons, protons, quarks, and so on—came, at least indirectly, from this explosion. Furthermore, the forces that today hold these particles together also emerged from it. Today there are four of these forces (they will be discussed in detail later); at that time there was only one—a unification of the four.

A question that immediately comes to mind regarding the explosion is: Where did the energy that produced it come from? A provocative new theory called "inflation" may provide the answer. Alan Guth of MIT was investigating some of the problems of the big bang theory in 1979 when he discovered that a sudden increase in the rate of expansion would solve many of them. This inflation lasted only from 10^{-36} to 10^{-34} second, but during this time the energy that would power the universe was created. Particles were generated in a tremendous reheating that followed inflation.

At the end of inflation the universe was still only the size of an orange. Less than a trillionth of a second later it had a radius of about three feet and a temperature of about 100 billion degrees. At this stage it was still composed mostly of particles. Quarks, the particles that make up protons and neutrons were abundant, but they were free at this stage. Other particles such as electrons were also present along with energetic radiation.

When the universe was a millionth of a second old the quarks began to clump together to form protons and neutrons. Soon, many of them began to decay, releasing radiation. As more and more radiation was released the universe became dominated by radiation. It was during this time that the first nuclei appeared: protons and neutrons began to clump together to form nuclei, first deuterium, then tritium, and finally helium. But that is as far as it went. And soon the universe became a boring place with little happening. It just continued to expand and cool. Then the giant cloud began to break up, and for a while the fragments continued to expand with the universe. Finally, though, they broke away and began contracting. Some became spiral in shape, others elliptical, and a few remained irregular. The first galaxies soon appeared. Then gigantic explosions forced the galaxies into clusters and clusters of clusters and the universe began to take on a mottled filamentary appearance. And within the galaxies massive stars began to cook the heavier elements in their thermonuclear furnaces. Carbon, oxygen, neon, magnesium, silicon, and iron formed, then were blown into space in supernova explosions. And from the debris came planets—then the first forms of life.

We have talked about how the universe began in considerable detail. But what evidence do we have that our theories are correct? We do, indeed, have evidence. One of the predictions of the big bang theory is that intense radiation was released from the expanding fireball early in its history. According to calculations this radiation expanded into the universe and cooled. We can, in fact, calculate how hot it should now be. Astronomers have shown that it should be about 3 K. This has been found to be correct. Also, astronomers have predicted that about 25% of the material that came out of the big bang should now be helium—and it is.

The story of creation is an intriguing one. It's like a detective story with surprises around every corner. But to understand it we must begin at the beginning. In the next chapter we will look at the evidence we now have for the existence of the "big bang."

CHAPTER 2

Discovery of the Expanding Universe

The story of creation begins with the discovery of the expansion of the universe. Looking at this expansion today we see that it does not involve our sun, our solar system, or even our galaxy, the Milky Way. None of them are expanding. It involves only the space between the galaxies. As we look at the billions of galactic systems beyond the Milky Way that dot the universe we see that each of them is racing away from us. Does this make us special? No, for upon closer examination we find that they are not just racing away from us; they are racing away from each other. It is the space between the galaxies that is expanding.

How do we know the universe is expanding? To answer this it is best to begin with our galaxy. The British astronomer William Herschel was the first to make a detailed study of it. As a young man Herschel was an oboist in the German regimental band, but at the Battle of Asterbrook in 1757 news reached him that the band members were going to be pressed into action. Shocked, he panicked and deserted; with his brother Jacob he fled to England, where he pursued his love of music and began giving lessons and composing.

But things soon changed. After reading a number of popular astronomy books his interest began to shift from music to telescopes and the stars. At first only a small amount of time was spent away from music, but eventually he turned almost entirely to astronomy. Interestingly, the change occurred when

Galaxy in Ursa Major. (Courtesy National Optical Astronomy Observatories.)

he was almost 40 years old. Before he was 35 he had barely looked at the stars. But from then on until his death he made good use of his time. It can be said that he truly opened the door to the heavens, first by developing techniques for building large telescopes, then by using them to make some of the most important discoveries ever made in astronomy.

Herschel contributed significantly to our understanding of the Milky Way galaxy. Using star counts he concluded that it was shaped like a huge disk. But he did not stop with our galaxy. Starting with a table of 103 nebulous objects published by the comet hunter Charles Messier, Herschel began a search for more. By the time of his death in 1822 he had catalogued a total of 2500. Noticing that most were situated in a direction away from the Milky Way, he believed that they were distant star systems, like ours, but outside it. At first he thought that all hazy objects were star systems of this type, but later he came to realize that a few were different and likely nearby.

Although William, and later his son John, catalogued these nebulous objects, they spent little time studying their structure. This was left to the Irish astronomer, William Parsons, 3rd earl of Rosse. When Rosse turned his giant 72-inch telescope toward the one in the constellation (group of stars) Canes Venatici in the mid-1800s he was amazed by its strange spiral structure. Then he discovered that many of the other nebulous objects also had a similar structure. Astronomers were intrigued with the new objects, but uncertain what they were. Rosse believed they might be clouds of stars, but others were inclined to believe they were gaseous.

As the year 1900 approached the controversy was still unresolved. It was now well established that most of these objects—particularly the spiral-shaped ones—were in a direction away from the Milky Way. Furthermore, there was now proof that a few of the nonspirals were gaseous.

The breakthrough that was finally to resolve the difficulty came in 1912. Miss Henrietta Leavitt of Harvard Observatory was studying two objects in the southern skies called the Magellanic clouds. Upon comparing exposures made several days apart she discovered that these clouds contained a large number of stars called Cepheids that varied periodically in brightness. Detailed study showed her that the larger, brighter Cepheids had a longer period; in other words they took longer to go from peak brightness to peak brightness. What did this mean? To Leavitt it implied that there was a relationship between period and brightness. But there was something different in this case: all of the stars in the Magellanic clouds were at roughly the same distance. This meant that the longer the period, the greater the true or "absolute" brightness of the Cepheid.

The significance of Leavitt's discovery was soon realized by

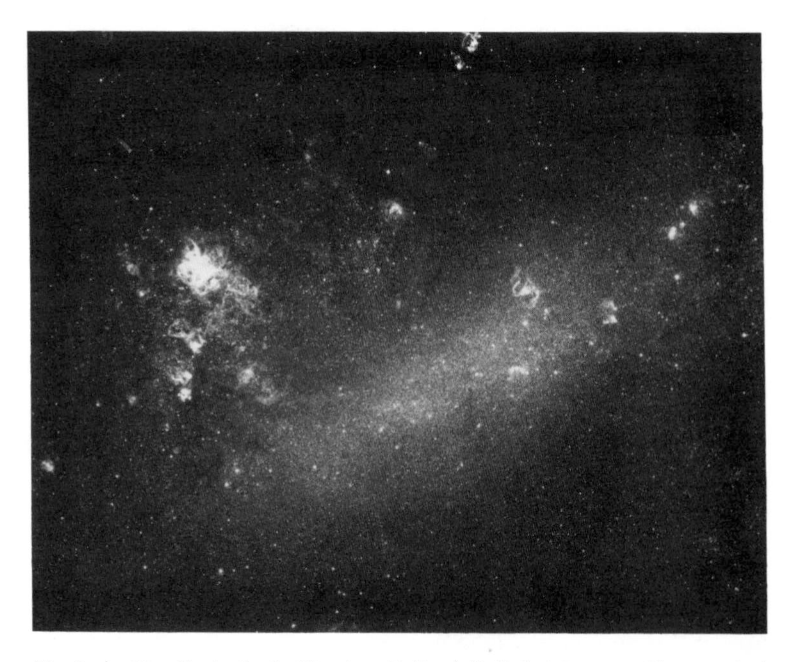

The Large Magellanic cloud. (Courtesy National Optical Astronomy Observatories.)

the Danish astronomer Ejnar Hertszprung and an American, Henry Norris Russell. If we knew the distance to one Cepheid, we could immediately determine the distance to any Cepheid by measuring its period and average brightness. This also meant that we could determine the distance to a cluster or group of stars that contained a Cepheid. But the problem of determining the distance to a Cepheid was not easily solved and several years passed before the solution came. Before we talk about the breakthrough that the solution brought, though, let us go back a few years and fill in some of the details.

In 1894 Percival Lowell, a rich, flamboyant Bostonian businessman, set up an observatory under the velvety black skies of Flagstaff, Arizona. He, too, had seen the strange, fuzzy objects in the sky—the nebulae—and although a few astronomers had already suggested that they might be clouds of stars, Lowell was not convinced. In fact, if he had believed they were stars he probably would not have bothered with them, as he had little interest in stars. He was interested in the planets—Mars, in particular. Indeed, it would be much more accurate to say that he was fascinated by Mars and the possibility that there might be an advanced civilization inhabiting it. In his zeal to learn more about it, he had ordered a spectroscope—an instrument that separates white light into its various colors.

Let's take a moment to look at the spectroscope. The figure shows that when a beam of white light passes through a prism it breaks up into a rainbow of colors. This happens because white light is composed of all colors; the prism merely separates them. It does this because light is composed of particles called photons that vibrate at a certain rate as they move through space. This vibrational rate is referred to as their frequency. Photons corresponding to red light, for example, have a different frequency than those corresponding to blue light. When a photon enters a prism or grating (the basic component of a spectroscope) its speed and direction depend on how fast it is vibrating. Because

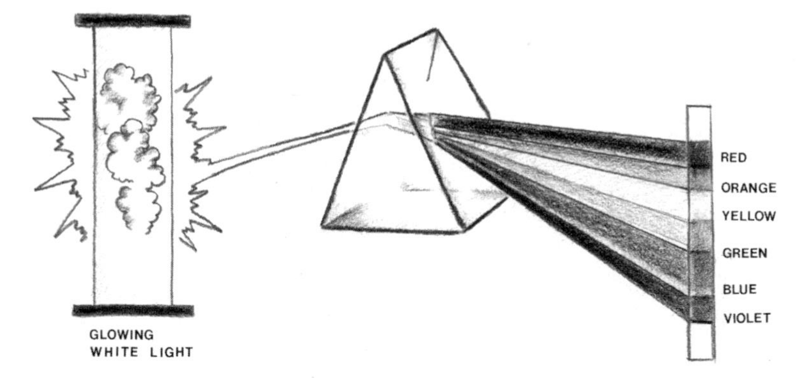

A simple spectroscope. Continuous spectrum.

of this, a beam of white light, which is composed of photons of many different frequencies, spreads out as it passes through the prism.

If, instead of white light, we shine the light from hot glowing hydrogen through our spectroscope, we do not get all the colors of the rainbow. We get a few sharp, brightly colored lines. And because a star is composed mostly of hydrogen we would expect similar lines (called spectral lines) from it. It turns out, though, that the lines we get come not from the star itself but from the atmosphere surrounding it (the light passes through this atmosphere). And because of this we end up getting dark lines (still in the same position) on a bright background. Anyway, the important point is that these lines are of tremendous value to the astronomer; they provide information about the star that could be obtained in no other way.

Now back to Lowell. In 1901 he received his spectroscope. Knowing that he lacked the patience or experience to work with such an instrument he hired a recent graduate of the University of Indiana, Vesto Slipher, to set it up and get it operating. Sli-

Spectral lines.

Vesto Slipher.

pher came to the observatory late in the summer of 1901. He was to be a short-term assistant, and was hired only as a favor to one of the other astronomers. It is perhaps ironic that he stayed 53 years. Indeed, he eventually succeeded Lowell as the director of the observatory.

Born on a farm in Indiana in 1875, Slipher was a reserved, cautious man who shunned publicity. But he was also the ideal man for the job; he had almost infinite patience and a considerable amount of mechanical expertise. Within a short time he had the spectroscope set up and soon was proficient in its use. For the first few years most of his attention was directed at the planets. Then in 1909 Lowell suggested that he try to obtain the spectrum of the spiral nebulae (at that time galaxies were called nebulae). Lowell believed that they might be solar systems in formation, and that their spectrum might tell us something about our system in its youth. Slipher was pessimistic. "I don't see much hope in getting their spectra . . . ," he said to Lowell.

But Lowell insisted that he try. Slipher began experimenting with various techniques and photographic films and soon discovered that his worst fear was not realized. The relatively small telescope that he was using (a 24-inch refractor) would not be a problem—at least not initially. What he needed most was a fast camera—and he soon got one.

In December 1910 he obtained his first spectrogram of the Great Nebula in Andromeda. It showed faint lines, not distinct enough to measure, but it was a start. This gave him confidence.

Throughout 1911 and 1912 he continued perfecting his technique, then in September 1912 he tried a 6-hour exposure of the Andromeda Nebula. This time the lines were much more distinct. But he was sure he could do even better. And indeed he did. On Novenber 15 and 16 and again on December 3 and 4 he obtained two more spectra that were even sharper. Finally, on December 28 he began an exposure that extended over three nights. He now had plates he could measure.

Then came the surprise. When he measured the lines he found they were all shifted from where they should have been. He was amazed and concerned—not because they were shifted, but because they were shifted by such a large amount. Astronomers were familiar with shifts of this type; ordinary stars exhibit them. They were considered to be due to the Doppler effect—a shift in frequency that occurs when a source of waves is approaching or receding. You have no doubt encountered the effect in relation to trains. The pitch of a train's whistle is higher when the train is approaching you than when it is receding. By measuring the amount that it has changed and comparing it to the frequency when the train is stationary, you can determine how fast the train is approaching or receding from you. The same thing can be done with stars. It therefore seemed as if it also applied to nebulae.

But according to Slipher's calculations the Andromeda Nebula was approaching us at the unheard-of speed of 300 kilometers per second. No star had ever been found with such a high speed. Slipher wondered about the result but eventually

The Andromeda galaxy. (Courtesy National Optical Astronomy Observatories.)

convinced himself that the shift had to represent a velocity. ". . . I cannot find any other explanation," he wrote.

He remeasured the Andromeda plates once more, then when he was sure that no mistakes had been made he wrote Lowell, who was now back in Boston, and told him about it. "It looks like you've made a great discovery," Lowell wrote back. "Try some other[s] . . . ."

Slipher then turned to an edge-on spiral in the constellation of Virgo. It also had a high velocity relative to Earth. In fact, it was three times higher than the Andromeda Nebula's velocity. To further complicate things it was moving away from us rather than approaching. He then turned to other nebulae and found that most of them were receding. By August 1915 he had measured a total of 15 and found that almost all were receding. "The striking preponderance of the positive sign indicates a general fleeing from us or the Milky Way," he said at the 1914 meeting of the American Astronomical Society at Evanston. He was careful, however, not to speculate on what the shift meant. Nevertheless, the crowd sensed that he had made a significant discovery, and he was given a standing ovation at the end of his talk.

What was Slipher's reaction to the high velocities? In the earliest stages of his work he was not convinced that the objects were galaxies. After the Orion Nebula was confirmed to be gaseous he wondered about the Andromeda Nebula. But by 1917 he had changed his mind and stated that the "island universe" theory was most appropriate for his observations. Even after he convinced himself that the objects were galaxies, though, he did not think in terms of an overall expansion of the universe. Early on he had noticed a group of galaxies in the same region of the sky that were all blueshifted, indicating they were approaching us; among them was the Andromeda galaxy. Opposite them in the sky were several redshifted galaxies. This convinced him that the velocities were due to *our* motion. In other words, our galaxy had a "drift velocity" relative to other galaxies. He clung to this idea throughout his career, but he did admit that it seemed strange that all the approaching ones were in the same small group.

By 1917 he had radial velocities (velocity along the line of sight) for 25 spiral galaxies. Most of his exposures had been over several nights—20 to 40 hours. But now he was getting to extremely faint galaxies—he was approaching the limit of the 24-inch telescope.

In the same year (1917) the Dutch astronomer Willem de Sitter announced that his theoretical studies were indicating that the universe may be expanding. This meant that all nebulae should be moving apart. But this seemed to have little effect on Slipher's observing program and he clung to his "drift hypothesis."

The redshifts that Slipher obtained were an enigma. Why did they occur? And the blueshifts were also confusing. What did it all mean? If you ignored the blueshifts there was the possibility of a velocity–distance relation for nebulae. And, indeed, several people looked into this, but none of them could establish a relationship.

Ludvik Silberstein decided to take a different route. He realized that the major difficulty centered around the distances. Distances to the galaxies were not known accurately. He therefore decided to concentrate on objects whose distances were known—globular clusters. And within a short time he found a velocity–distance relation—or at least thought he did. Knut Lundmark of Sweden was skeptical; he pointed out that globular clusters were quite different from nebulae—they only had a typical radial velocity of 31 kilometers per second compared to 800 for nebulae. Then when he tried to verify Silberstein's results he found he could not.

For a while confusion reigned. Then came Edwin Hubble. Picking up where Slipher left off he probed the universe to its outer limits, and in so doing made one of the most important discoveries ever made: the universe is expanding. His discoveries threw the world of astronomy into confusion, and for a

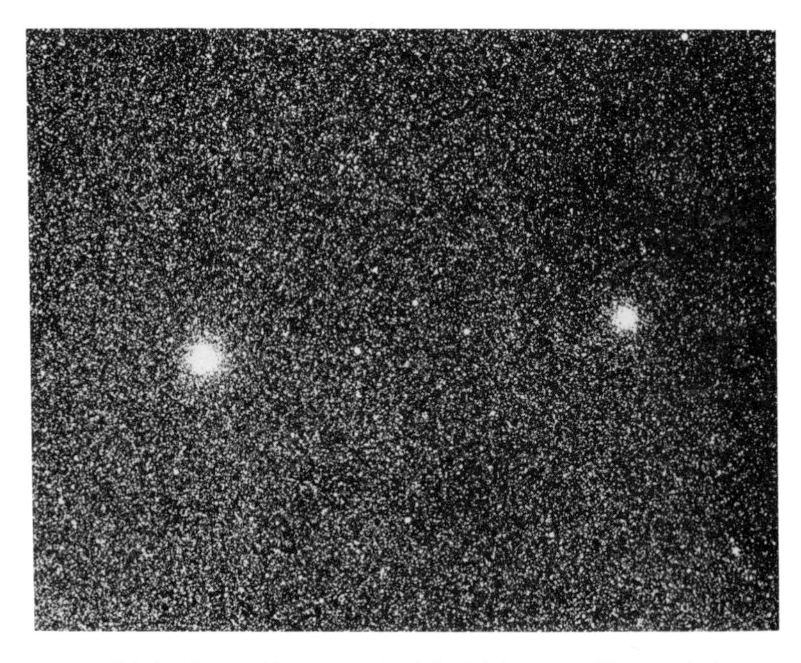

Globular clusters. (Courtesy National Optical Astronomy Observatories.)

while debates raged. But Hubble was a careful and meticulous observer and when the evidence was finally in there was no doubt that he was right.

HUBBLE

Hubble was not just a dedicated and capable scientist. He was also a Rhodes scholar, a decorated war hero, and an outstanding athlete who played on a championship basketball team and boxed the light heavyweight champion of Europe. Most people who worked and associated with him agreed that he was a great astronomer—one of the greatest who ever lived. But some of them found him cold, aloof, and even arrogant. Shapley, an astronomer who worked with him for many years, wrote, "Hubble just didn't like people. He didn't associate with them, didn't like to work with them . . . . Hubble and I did not visit very much. He was a Rhodes scholar, and he didn't live it

Edwin Hubble and sister. (Courtesy Henry E. Huntington Library.)

down. He spoke with a thick Oxford accent . . . . The ladies he associated with enjoyed that Oxford touch very much. 'Bah Jove,' he would say." But Shapley admitted, perhaps reluctantly, that he was "picturesque."

Others, however, found him to be charming, personable, and always eager to help. His student, astronomer George Abell, wrote of him, "He had a certain nobility about him; he was a gentleman in the true sense of the word. Some people have described him as aloof, but I found him to be warm and kind, and I found that he always had time to talk with anyone who sought his advice or help." And his colleague N. U. Mayall wrote, "To have known Hubble as a friend and neighboring senior colleague is an experience [I] will long cherish. . . . [He] was always helpful and encouraging."

Hubble was born in Marchfield, Missouri, in 1889. Interestingly, one of his earliest memories was of astronomy. He was asked by his parents what he wanted for his fourth birthday. They were surprised when he asked, not for a toy, or even a trip to the circus, but rather that he could stay up that night to watch a meteor shower that was expected.

When he was nine the family moved from Marchfield to nearby Evanston. His memories of Marchfield were pleasant ones, and interestingly, Marchfield never forgot him. In 1952 he received a telegram from the mayor of Marchfield requesting his presence as honorary marshal in the next July 4th parade. He would be given a medal and citation. He yearned to attend but was unable due to heart problems. He died shortly after the parade took place.

In high school he was a star athlete, participating in both football and track. But he was also an outstanding scholar who managed to be outstanding without studying. This was a slight embarrassment to him, though, when he walked up on the stage to get his diploma. The principal studied him for several moments before all the students, their families and friends, then said, "Edwin Hubble, I have watched you for four years and have never seen you study for ten minutes." He stood there silently for what seemed hours to Hubble, then continued, "Here is a scholarship to the University of Chicago."

He had dreamed all through high school of the day when he would go to college and play for their football team. Football was his first love and his high school record was outstanding. On the day he left for college, however, his mother asked him to forgo football. It was too rough and he might get hurt. He was dumbfounded, but his arguments fell on deaf ears. She had made up her mind and wanted him to promise her he would not play. Reluctantly he made the promise, the hardest one he ever made in his life. He later said, "Nobody but myself could understand what this renunciation meant to me."

Perhaps the strangest part is that he was not forbidden from participating in boxing. And indeed he boxed throughout his university career, both at the University of Chicago and later at Oxford. His boxing skills were great enough, in fact, that they came to the attention of a Chicago promoter who tried to get him to turn professional and train to fight the world heavyweight champion, Jack Johnson. He declined the offer. Later, however, he fought the light heavyweight champion of Europe, Georges Carpentier, in an exhibition fight and managed to get a draw.

To help support himself during his college days he worked during the summers as a surveyor and engineer in the region around the Great Lakes. The area was young and most of the nearby towns were loaded with "tough characters." This aspect never bothered Hubble. He was walking through the railroad yard of one of these towns one evening when two ruffians held him up with a knife. He looked at the knife, then at them, and began to laugh. Ignoring their demands he walked past them and continued on his way. One of the men ran up and stabbed him in the back just below the shoulder blades. Hubble turned and struck him so hard he knocked him unconscious. Hubble then went for the other man but he was already sprinting away.

He leaned down and inspected the unconscious man, making sure he was alive, then went and got medical attention for his knife wound.

In the fall of 1910 he went to Oxford, where he studied law for three years. During the summer he spent a considerable amount of time touring Europe. It was during one of these visits that he got himself into a duel over a woman. He related the tale to his wife years later. "The day I got to Kiel," he said, "I went for a swim. A woman swimming a little ways off threw up her hands and called for help. I swam over and brought her to shore. She was none the worse, so I left her with some of her friends and walked away." The woman, who was about 30 and, according to Hubble, "very pretty," was the wife of an important German naval officer with a title. The next day the officer located Hubble and thanked him. They soon became good friends, playing tennis and swimming together. The officer's wife never came with them but she was, of course, always in the house whenever he was there.

A rumor started among the officers that Hubble had made a pass at the officer's wife. Hubble denied it, saying that he was polite to her, but that was all. Nevertheless, whenever such an incident occurred the husband was forced to challenge the wrongdoer to a duel. "It was a surprise to me," Hubble said later, "when he called at my room and said, with some embarrassment, that my conduct towards his wife had made it necessary for him to challenge me to a duel."

Hubble was dumbfounded, but agreed. He had heard that such duels were fought, but never believed he would be involved in one. The following morning he went to the officer's home where he was led to a large oak-lined room. The officer gave him the choice of two pistols, then both men moved to opposite ends of the room. After a short count they raised their pistols and fired. Hubble could not bring himself to shoot directly at the man, who he still felt was his friend. He shot wide into the oak wall, and luckily the officer did the same thing so neither man was hurt. They both bowed, then with no further words Hubble left the house. And soon afterwards he left Kiel. He felt regret at the incident, and knew that his friend had been forced into it by circumstances. He felt that they parted friends but he never saw him again.

In the summer of 1913 he sailed back to the United States, passed the bar exam on September 2, and began practicing shortly thereafter in Louisville, Kentucky. His law practice flourished and he made a considerable amount of money, but the love for astronomy that had developed many years earlier began to tug at him. Second thoughts about his life and ultimate aims occurred more and more often. About a year later, after some deep soul-searching, he made a decision: it was astronomy he was really interested in, and he would never be satisfied unless he pursued this urge. "Astronomy is something like the ministry," he said later. "No one should go into it without a 'call.' I got that unmistakable call, and I knew that even if I were second rate or third rate, it was astronomy that mattered."

In 1914 he took down his law shingle and registered at the graduate school of the University of Chicago to work on a Ph.D. in astronomy.

Much of his time during his graduate years was spent at the Yerkes Observatory at Williams Bay, Wisconsin, on Lake Geneva, studying "nebulae." There was still considerable controversy about these small cloudlike patches of light, but Hubble eventually became convinced that they were "island universes" of stars, distinct from our Milky Way. His thesis was titled, "A Photographic Investigation of Faint Nebulae."

He was nearing completion of his thesis when Dr. Hale of Mt. Wilson Observatory in California visited Yerkes and offered him a job. Hubble was delighted. But fate intervened. The United States entered World War I. Hubble stayed up all night finishing his thesis the night before his orals. The next day he took them, passed, then rushed out and enlisted in the army. He telegraphed Hale, "Regret, cannot accept your offer. Am off to war."

Hubble rose quickly through the ranks, eventually becoming a major. With only days to go before the end of the war a shell burst next to him, knocking him unconscious. The next thing he remembered was waking up on a cot in a small tent with an extreme headache. No one else was in the tent. He quickly felt both arms and legs to see if they were intact. They seemed to be okay, so he dressed himself and walked out of the tent.

He was, however, more seriously wounded than he thought. He had a concussion and an injured right elbow, and several cuts over his body. He was, in fact, never completely able to straighten his right arm after that, and it put a hook in his golf game. Rather than work to get rid of it he eventually gave up golf.

Late in the summer of 1919 he returned to the United States and went immediately to Mt. Wilson Observatory. Hale had earlier asked him to come as soon as the war was over.

THE VELOCITY–DISTANCE RELATION

Hubble spent his first years at Mt. Wilson working on the problem that had challenged him years earlier as a graduate student: What are the nebulae? Even then he was convinced that they were "island universes" of stars. Now, with the hundred-inch Mt. Wilson reflector he set out to prove it. He selected several nearby nebulae for detailed study; among them were the Great Nebula of Andromeda and the spiral nebula in Triangulum. Long exposures of these nebulae had already shown "condensations," but astronomers were not convinced that these regions were stars. Hubble directed the hundred-inch telescope at several of them and took long exposures—sometimes extending over several nights. Then, looking carefully at their outer regions, he found stars. With persistence and patience he obtained photograph after photograph showing the outer regions of nebulae resolved into stars. There was now no doubt: the spiral nebulae were composed of stars.

But did this mean they were necessarily outside the Milky Way? No, it was still possible that they were within it. Hubble needed something more—and he soon found it. Some of the stars he resolved were variable; in particular, they were Cepheid variables. Using the Cepheid period–luminosity relation he could determine their distance merely by determining their period. And soon he had the result he wanted: the nebulae were outside the Milky Way. The Andromeda Nebula, for example, was 800,000 light-years away (later adjustments gave 2 million light-years). This was far greater than the size of the Milky Way.

With this out of the way he turned his attention to the velocities and distances of the nebulae. He was quite familiar with Slipher's work and with Silberstein's effort to find a velocity–distance relation. He also realized that the major problem was the distance to these objects. Without accurate distances it would be impossible to prove a relationship. The first step, then, was to determine the distances of the nebulae that Slipher had obtained redshifts for.

Hubble started with a total of 46 objects with known redshifts. Most had been measured by Slipher, but several had been verified and a few added by Hubble and his jovial, easygoing assistant Milton Humason. For the nearest nebulae there were few problems; Cepheids could be seen and their distances could be obtained relatively accurately. With the distances to several of the nearby ones determined, Hubble began looking at the brighter stars in them. They all seemed to have approximately the same magnitude (brightness) so he decided to use them as "standards." With this and the knowledge of how their light fell off with distance he could use similar stars in dimmer nebulae to determine their distance. Finally the nebulae themselves were used. To a first approximation nebulae are all about the same brightness; thus, if he knew the brightness and distance of a nearby one he could use them to determine the distance to dimmer ones.

This allowed him to determine distances for 24 of the 46 objects with known redshifts. He corrected each of them careful-

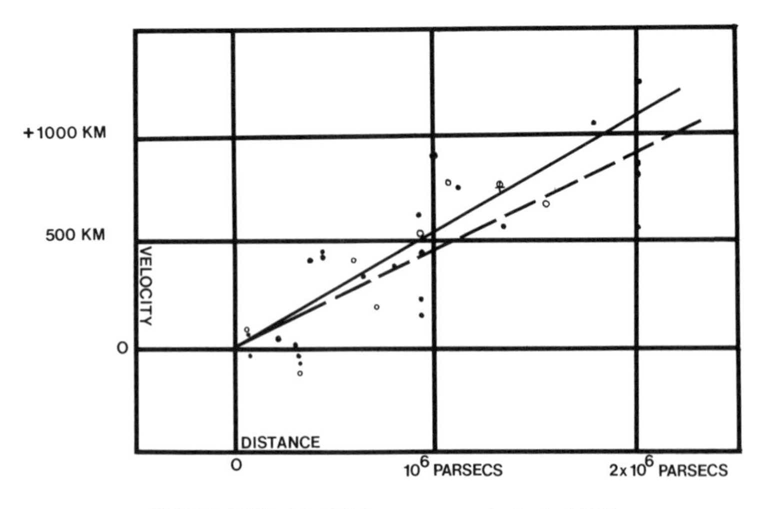

Hubble's 1929 plot of distance versus velocity (redshift).

ly for the motion of our sun, then plotted velocity against distance. The scatter in the points was considerable (see diagram) but a relationship was visible. He drew a straight line roughly through the middle of the points.

In January of 1929 he published his results. He began cautiously with the statement, "The present paper is a reexamination of the question [velocity–distance relation], based on only those nebular distances which are believed to be fairly accurate." With such a statement he was obviously trying to overcome the skepticism that had developed since Silberstein's earlier attempt. He then went on to talk about his distance determinations. "The first seven distances are the most reliable, depending . . . upon extensive investigation of many stars . . . . The next thirteen distances are subject to considerable probable error but are believed to be the most reliable values at present available."

There is no question, and it is quite evident in the paper, that Hubble had been influenced by a prediction made a few years earlier by the Dutch astronomer de Sitter that there should be a cosmological redshift. Toward the end of the paper he stated, "The outstanding feature . . . is the possibility that the velocity–distance relation may represent the de Sitter effect. . . ."

Hubble's paper was barely out when Shapley wrote a brief note to the same journal criticizing it. His main concern was the use of the magnitude of a nebula as an indication of its distance. He felt that not enough was known about nebulae to go this far, and that caution should be used in jumping to any conclusions. It is well known, of course, that Shapley did not care for Hubble personally; in addition, he had also pointed out several years earlier that there might be a relation between velocity and distance of nebulae. There is little doubt that Shapley was unhappy to see Hubble making the very discovery he had predicted. His dislike for Hubble is further evidenced in that he said that Hubble was using his methods for determining the distances to nebulae without acknowledging them.

Hubble did, indeed, fail to mention Shapley; even worse he did not mention Slipher in his 1929 paper (despite the fact that most of the redshifts were due to Slipher). Both of these oversights were corrected, however, in his 1931 paper.

Hubble realized at this stage that his case was far from airtight. He would have to extend his graph to nebulae much deeper in space. With this in mind he gave Humason a list of faint, presumably more distant, nebulae for which he wanted redshifts. He then turned to his distance measurements. They would have to be improved.

Within two years Humason had obtained the spectra of 37 more nebulae and F. G. Pease, also of Mt. Wilson, had obtained an additional nine. The most distant objects were now 16 times farther out than those of his 1929 plot. Hubble had now gone carefully over his "cosmic distance ladder" strengthening each rung. He realized that each rung depended on the one below it, and if something was wrong with a lower one the entire scale would be thrown off. He identified several types of objects—Cepheids, novae, irregular variables—in ten of the closer nebu-

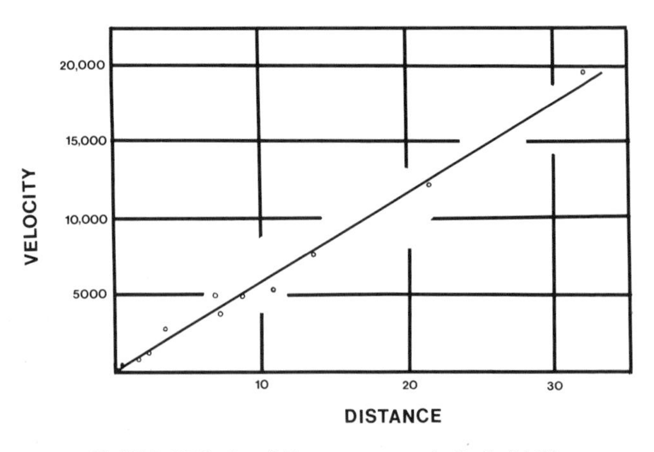

Hubble's 1931 plot of distance versus velocity (redshift).

lae. The Cepheid variables were, of course, the most valuable. He then carefully studied the brighter stars in 40 nebulae and showed that their magnitudes were independent of the brightness of the nebulae themselves.

With a new and more accurate distance scale and many new velocities Hubble was ready to extend his plot of velocity versus distance. Rather than plot individual velocities this time, however, he calculated the average value for groups (most nebulae reside in groups) and used them. His previous points were now all in the lower right-hand corner of the graph. When the plot was finally made there was little doubt. The scatter of the points was much less this time as compared to his earlier plot. It would now be difficult for anyone to argue that there was no relationship between velocity and distance. But Hubble was still cautious. He was now less willing to interpret his redshifts as velocities than he had been in 1929. He preferred to refer to them as "apparent velocity displacements."

The reason for Hubble's change in attitude was the rather sudden change that had occurred in theoretical cosmology. De Sitter's model had generally been ruled out—it predicted a universe with no mass in it and this was in contradiction to observation. Furthermore, another model had been put forward by a Belgian, Georges Lemaître. It is not known how well Hubble understood the details of these models, but they did have some influence on him. The development that had the most effect, though, was a suggestion made by a co-worker at Mt. Wilson, Fritz Zwicky. Zwicky was not convinced that the redshifts were due to velocity; he thought that they might be due to an interaction between light and the matter in space—a kind of "tiring" of light. According to his ideas the interaction might cause the light to lose energy and become redder, and as a result the spectral lines would shift. Zwicky was drawn to this conclusion because of the exceedingly large velocities Humason was now obtaining—up to 10% the velocity of light. Zwicky was sure this was impossible. Although he did not work out the details of the theory, the suggestion itself was enough to worry Hubble. He decided it was best at this stage to concern himself only with the observations.

In the years after 1931 Hubble and Humason continued to extend their graph to more and more distant nebulae. In 1934 they published the velocities of 35 more, and in 1936 they added a further 100.

INTERPRETATION

Hubble struggled with the interpretation of the redshift for years. That it should not necessarily be treated as a velocity was evident in his use of the words, "apparent velocities." In a 1931 letter to de Sitter he said, "We use the term 'apparent' velocities in order to emphasize the empirical features of the correlation. The interpretation, we feel, should be left to you and very few others who are competent to discuss the matter with authority."

But Hubble was not altogether uninterested in the interpretation. He knew it was important to resolve the difficulty, and in an effort to get at the roots of the problem he turned to theoretical cosmologist Richard Tolman. Tolman was already working on several important cosmological problems, and in 1934 had published a book titled *Relativity, Thermodynamics and Cosmology* that eventually became a classic in the area.

One of the reasons Hubble selected Tolman to collaborate with him was no doubt his proximity. Tolman was associated with nearby Caltech. Furthermore, they were already cooperating in other matters related to the building of a larger telescope. But this is, of course, not the only reason. Tolman was, at the time, one of the foremost authorities in the field.

What the two men hoped to do was set up several tests that could be used to decide between the various theoretical models. These tests would, presumably, involve observations of such things as the brightnesses of nebulae, and the number of nebulae at various distances. In an effort to make the results definitive they selected two models on which to base their determination. One was a standard expanding universe obeying the relativistic laws of gravity; the other was a static model that had been put forward by Einstein many years earlier.

Their major hope was to determine between a true recessional interpretation of the redshift and the "tired light" theory of Zwicky. They were, unfortunately, unable to come to any firm conclusion. Their work did, however, set the stage for later efforts, and in this sense it was a success. Although there appeared to be little observational proof, Hubble and Tolman both admitted they were inclined toward the interpretation that the redshift did, indeed, indicate that the nebulae were receding from us.

Their joint paper appeared in *The Astrophysical Journal* in 1935. Although it did not have a solid case against Zwicky's theory, others soon showed that there were serious problems and it eventually fell out of favor.

Hubble was now at the limit of the hundred-inch telescope. To go further would require a larger telescope. But he had established that there was a relationship between velocity and distance, and within a few years it was generally accepted that the universe was, indeed, expanding. Hubble summarized his results in 1936 in his book *Realm of the Nebulae.*

CHAPTER 3

Cosmology of the Mind

Cosmology, like all sciences, consists of two branches: observation (experiment) and theory. And for the most part these two branches complement one another; in other words they act as a team. When a new observation is made, for example, theoreticians scramble to make it fit the theory. If it does not fit, the theory is discarded and a new one is devised. Although this is the way things happen today, it was not always so. In the early part of this century, theory and observation progressed largely independent of one another. Not until 1929, when Hubble announced his discovery of the redshift–distance relation, did things begin to change.

When Hubble made his announcement, theoreticians in Europe had already been considering theories of the universe for over ten years. Before we look at the details of these theories, though, let us consider what we would expect such a theory to contain.

One of the first questions we are likely to ask is: Does the universe have a boundary? And if so, what is it like? This is an intriguing and, needless to say, mind-boggling question. There are obviously two possibilities: it does or it does not. The problem is that either answer leads to difficulties—severe ones. Consider the case where it does not. Clearly, if this is so, the universe must extend to infinity. But astronomers do not like the concept of infinity. After all, what does it really mean? It's certainly difficult, if not impossible, to visualize. The alternative, however, leads to just as many problems. If the universe does have a boundary we immediately ask: what is on the other side?

Sir Isaac Newton was the first to consider these perplexing questions. He wanted a mathematical model of the universe that would not only overcome the boundary problem, but would also allow him to make predictions. He started out by considering a finite universe uniformly distributed with stars but soon realized it would be unstable. According to his law of gravity every object in the universe is attracted to every other object, and such a universe would therefore soon collapse inward on itself. He also convinced himself that something similar would happen if he had a finite amount of mass in an infinite universe. He wrote, ". . . if the matter was evenly distributed throughout an infinite space it would never convene into one mass; but some of it would convene into another mass and some into another, so as to make an infinite number of great masses scattered at great distances from one another throughout all that infinite space."

This is an interesting statement in that we know that the universe is indeed populated by great masses—galaxies—scattered at great distances from one another. In Newton's day, of course, the existence of galaxies was not known. But aside from this it is interesting that Newton did not see a way around his problem. First of all, if the universe was rotating, the matter would not necessarily collapse inward. And furthermore, we can, as Einstein did, think of the galaxies as molecules of a gas. In this case, if the "effective temperature" of the "galactic gas" is greater than zero, they will not fall into one another, just as the molecules of a gas with a temperature above zero do not fall into one another.

Newton's major stumbling block, however, was that so little was known about the universe. The solar system was fairly well understood, but astronomers had only a vague notion about the makeup of the stars. Furthermore, they had no idea how far away they were. The obstacles that Newton had to overcome were, to say the least, overwhelming. The attempt was just ahead of its time. And he must have eventually realized it, for he gave up in disgust.

In time, though, the approximate distances of some of the nearby stars became known. And to the surprise of astronomers they were much farther away than expected. The universe was huge! Armed with this information the Englishman Thomas Wright proposed in 1750 that we live in a group of millions of stars shaped in the form of a grinding wheel. If you look along the plane of the "grinding wheel," he said, you see many stars. And, indeed, we know we do—we see the Milky Way. Furthermore, he pointed out that if you look in any other direction you see few stars. And again we know this is the case. This, it turned out, was one of Wright's better ideas; later on he developed some that were quite weird. Anyway, it got the ball rolling, for in 1775 Immanuel Kant proposed that the Milky Way was only one of many similar systems—"island universes" of stars—and that these systems extended off in space to infinity. The idea was tossed around, talked about, and argued about for decades. But eventually, as astronomy progressed, and our knowledge increased, astronomers came to realize that Kant was right.

EINSTEIN'S UNIVERSE

Newton may have made the first attempt to formulate a mathematical model of the universe, but it was Einstein who published what can be called the first modern cosmology. About a year after he completed his general theory of relativity (1916) he published a paper titled, "Cosmological Considerations on the General Theory of Relativity." And just as Newton had based his cosmology on his theory of gravity, so too did Einstein base his on general relativity. Yes, general relativity is also a theory of gravity, but it is quite different from Newton's theory. In Newton's theory gravity is considered to be an action-at-a-distance force; this means that if something suddenly happens to change the gravitational field of, say, the Earth, the moon

Albert Einstein. (Courtesy AIP Niels Bohr Library.)

immediately compensates for it. In Einstein's theory, on the other hand, gravity is thought of as a "curvature" of space. According to Einstein, matter (for example, the sun) curves space, and other matter (for example, the planets) move in a "natural" manner through this curved space. By natural I mean "following the curve."

Einstein's route to relativity was not an easy one. His early career can only be described as catastrophic. Even he referred to it as a "comedy." Everything he did seemed to fail. He did not complete his studies in the gymnasium in Munich, he failed the entrance exam to the ETH in Zurich, and the first manuscript he submitted as a Ph.D. thesis was rejected. It might seem, with so many academic failures, that he was a poor student. Not so. Even as a youth he got excellent grades and was usually at the top of his class. His mother wrote, "Yesterday Albert received his grades, he was again number one, his report card was brilliant." His problem, perhaps, if you can call it a problem, was that he was fiercely independent, and most of the time preferred self-study to attending class.

But he was, without a doubt, a talented person. Where did he get this talent (perhaps I should call it genius)? It's hard to say. His father—a kindhearted, easygoing, but passive person who seemed to go from one financial failure to another—showed some mathematical ability as a youth. But it never developed because his family was too poor to send him to university. So it seems likely that Albert's mathematical ability came from his father. His love of music, on the other hand, no doubt came from his mother, who was an excellent pianist. He took violin lessons from age 6 to about 13. He particularly loved Mozart, but was also partial to Bach and Beethoven.

As a youth Einstein generally kept to himself. He hated sports, although later in life he had a sailboat and loved to sail on Lake Caputh. Much of his time was spent reading and studying. He particularly liked popular science books. The one person he was close to during this time was his sister Maja. He was, in fact, especially close to her throughout his life. ". . .I miss her more than I can easily explain," Einstein wrote shortly after her death.

He began studying calculus on his own when he was about 12 and continued until he had mastered it at about 16. At college, although a fairly good student, he frequently skipped classes. This may be why he was not looked upon with favor by several of his professors. Of the four others who graduated with him, three were taken on immediately at ETH. Weber, his physics teacher, considered him briefly, then passed him over for two mechanical engineers. "I was abandoned by everyone, starting at a loss on the threshold of life," he said later, referring to the time just after graduation.

It was the low point of Einstein's life. He had no job; he wanted to get married to a classmate, Mileva Maric, but his parents were strongly against it. Furthermore, his father's business had just failed. He wrote, "What oppresses me most . . . is the [financial] misfortune of my poor parents. Also it grieves me deeply that I, a grown man, have to stand idly by, unable to do the least thing to help. I am nothing but a burden to my family . . . . Really, it would have been better if I had never been born."

To make things even worse, his father died shortly afterward of a heart attack. Einstein visited him in the hospital, but when his father knew the end was near he asked everyone to leave. Einstein did not want to leave, but went along with his father's wish. He said later that he felt guilty about this the rest of his life. The one bright spot was that his father consented to his marriage just before he died. And although his mother never did like Mileva, she went along with it. He was married shortly thereafter, but as it turned out, it was not a happy marriage. They were divorced in 1918.

Things finally took a turn for the better with an offer of a job as a patent officer in Bern in 1901. And soon he had made his first major contribution to physics—the special theory of relativity. From then on there was no turning back.

Einstein eventually became world-famous, but he remained unassuming and humble throughout his life. Asked once if he would do it all over again he replied, "No, I'd be a plumber." On another occasion he said that a lighthouse keeper would be an ideal job for a theoretical physicist. And several times he told would-be scientists to earn their living at a nondemanding job such as a cobbler, so as to avoid the "publish or perish" pressure that undermines the joy in creative work.

Einstein, from the very beginning, had intended that his theory be universal. He wanted it to apply not only to the solar system and stars, but also to the entire universe. But if it was to do this he believed that it had to incorporate a principle that had been put forward a few years earlier by the Austrian physicist Ernst Mach. Mach, who had obtained his Ph.D. from the University of Vienna in 1860, had made major contributions to physics in his youth, but in later years had become an outcast in the physics community because of his strange views, the most radical of which was his disbelief in the existence of atoms. "If atoms exist, show me one," he would shout at his critics.

Einstein was introduced to Mach's book *Science of Mechanics* while a student at Zurich. "It had a profound influence on me," he said later. What influenced him most was a principle stating that the properties of space had no independent existence, but were dependent on the mass and its distribution within it. Rotation was only rotation with respect to the fixed stars, according to Mach. "There is no such thing as absolute rotation," he said. But Newton had given an excellent argument for the absoluteness of rotation. As evidence, he pointed to the curved surface on the water in a pail when it was swung around. Mach was not convinced. He admitted there was a centrifugal force with respect to the stars. But take away the stars and, as far as he was concerned, the force went with it.

Einstein visited Mach in 1911 to talk about space, time, and no doubt, atomic theory. At the time Mach was crippled with arthritis and well into his seventies. Furthermore, he was almost deaf and Einstein had to shout to be heard. Although Mach had earlier leaned favorably toward relativity he was at this stage strongly against it, but apparently he kept his views to himself. He had planned on writing a book pointing out what he thought were flaws, but died before completing it. He did not, however, keep his views on atomic theory to himself and Einstein no doubt gave up trying to convince him. Anyway, they parted friends. The following year, the same year in which Einstein published his general theory of relativity, Mach died. Einstein later said of him, "[He] was a good mechanician, but a deplorable philosopher."

Einstein began his cosmological paper of 1917 by looking at the difficulties Newton had encountered. Following Newton he considered an infinite universe uniformly filled with matter but found, as Newton had years earlier, that it was unstable. And because Einstein was firm in his conviction that the universe was static, he quickly ruled it out. He considered a finite amount of matter in an infinite universe but found it also gave problems.

Not only did the universe seem to be unstable but also there was a serious problem with the boundary. Furthermore, as I mentioned earlier, Einstein wanted to incorporate Mach's principle. The only way he could do these things, it seemed, was to modify his equations. In his paper he began: "I shall conduct the reader over the road that I myself travelled, rather a rough and winding road, because otherwise I cannot hope that he will take much interest in the result at the end of the journey." He then went on to say that he was going to make a "slight modification" of his equations. This modification amounted to the addition of a term that came to be known as the cosmological constant. Einstein felt that it destroyed the beauty of his equations by making them more complex, and he was reluctant to change them, but saw no other way. He realized that stars moved, but was convinced that, on the average, these motions canceled out and the overall universe was static. With his original equations he could not get such a universe.

Incidentally, in setting up his theory, Einstein had to make two further assumptions. The first was that the universe was homogeneous; in other words it was the same everywhere. And second, he assumed that regardless of where you were, the universe appeared the same in all directions. These two assumptions remain in all cosmologies today.

Now, let us take a moment to look at the cosmological constant. What exactly was it? First of all, it was a repulsion term that countered gravity. Gravity, as I am sure you know, pulls masses together; the cosmological term pushed them apart. But there was a problem. It was well known that Einstein's equations were highly accurate over short distances (for example, within the solar system). This meant that the cosmological constant could be important only over extremely long distances. And, indeed, this was the case.

With his new equations Einstein got a spherical universe.

And, with ingenuity, he got around the problem of the boundary. According to his general theory of relativity, matter curved space. Indeed, Einstein had already shown that he could derive the orbits of the planets around the sun using the curved space concept. This meant that if space was strewn with matter (i.e. stars), it should have an overall curvature. Therefore, if you started out traveling in a particular direction, you would trace out a huge circle and eventually arrive back at the same point. In short, the universe was finite, had no boundary, no center, and there was no favored place in it. It was like the surface of a sphere. We refer to such a universe as "closed."

And if the universe was, indeed, closed you could calculate its circumference; all you needed to know was how much matter was in it. A first approximation to this number was worked out by Hubble a few years later, and a value of 10^{11} light-years was obtained—easily big enough to satisfy anyone.

Earlier I said that Einstein's universe was spherical, but, depending on how you looked at it, it could also be cylindrical. If you drew it in the way most scientists did, you got a cylinder. The three dimensions of space were tied up in the curved part of the cylinder, with time along its axis. The Dutch astronomer de Sitter later pointed out, though, that the way Einstein treated space and time was flawed. They should be treated as two parts of a space-time continuum; Einstein did not do this. He kept time distinct.

In retrospect, there is a strange irony in Einstein's theory. Einstein's main purpose in introducing his cosmological constant was to keep the universe static. And indeed it did. But a few years after he presented it the British astrophysicist Arthur Eddington showed that it was "barely" static. Sure, it was in a state of equilibrium, but it was an unstable equilibrium. It was finely balanced between expansion and collapse—like a pencil balanced on the blade of a knife. A slight shove in one direction and it would expand; a slight shove in the other direction and it would collapse.

Willem de Sitter.

DE SITTER'S UNIVERSE

As we just saw, the matter in Einstein's universe played an important role. It curved space and allowed him to get around the problem of a boundary. You can well imagine, then, his dismay when shortly after he published his theory he heard that a Dutch astronomer, Willem de Sitter, had discovered a solution he had missed. And de Sitter's universe had no matter in it!

Born in Holland in 1872, de Sitter studied astronomy at the University of Groningen. When he was nearing completion of his Ph.D. he met Sir David Gill, the director of the observatory at the Cape of Good Hope in South Africa, who offered him a position. And in 1897 he left for the Cape, remaining there for two years. He returned to Groningen in 1899, and a few years later became professor of astronomy at the University of Leiden. He was immediately drawn to Einstein's general theory of relativity and was partially responsible for the widespread interest in it. Upon receiving a copy of the theory from Einstein he passed it on to Eddington, who soon became fascinated with it.

But how could de Sitter's model of the universe be taken seriously? After all, we know that the universe has matter in it. de Sitter answered this by saying, "The universe is mostly empty anyway, so what's wrong with a cosmos that is all space?"

de Sitter published his paper in the same year Einstein did (1917). Upon examining Einstein's revised equations he found a solution that Einstein had somehow overlooked. Furthermore, he did something that Einstein had not: he derived observational tests for both his and Einstein's theories and checked them. In addition, he pointed out a number of problems in Einstein's model, but realized that his own model had at least as many. The major problem with his theory was that particles and light rays acted strangely in it. In particular, it predicted a shift of the spectral lines for distant objects toward the red end of the spectrum. This was due, according to de Sitter, to a slowing of time at large distances. Clocks at these distances would appear to run slow compared to clocks nearby. This was eventually referred to as the "de Sitter effect."

It was well known at the time that receding objects gave spectra with redshifted lines. Was the de Sitter effect due to receding objects? de Sitter looked into this, first in relation to stars, which he found had velocities too low to give significant shifts. Then he considered nebulae (although nebulae at that time were still not well understood). He was not familiar with Slipher's results directly and referred only to a few that had been quoted by Eddington. Among them was the large blueshift for the Andromeda Nebula, which indicated it was approaching. Several of the others, however, indicated recession so de

Sitter was not put off. He indicated that there appeared to be some agreement with his predicted redshift, but that more measurements would be needed. He was still uncertain at this stage, though, as to whether the redshift indicated motion, even going so far as referring to it as a "spurious red shift."

It was not until later that Hermann Weyl and then Eddington took the redshift as an indication of the expansion of the universe. Although de Sitter's universe was empty, Weyl showed that if you placed two masses in it they would separate from one another. This meant that all masses would "scatter" from one another, and that the universe should be expanding. This prediction was actually made several years before Hubble made his announcement.

Interest in the de Sitter and Einstein universes continued for over ten years. They were talked about and examined in detail, but both were flawed. Einstein's model was discarded soon after Hubble made his announcment. It was a static model, and there was no way it could explain expansion. The major problem with de Sitter's model was the one that had been known from the beginning—it contained no matter. And eventually this led to its downfall.

THE YEAR 1930

Einstein clung to his model throughout the 1920's, despite his disenchantment with the cosmological constant. Then in 1930 he visited Hubble at Mt. Wilson Observatory. He talked with him, toured the observatory, and looked at nebulae through the hundred-inch reflector. Soon afterwards he made his announcement: He was abandoning the cosmological constant. And he kept his word. Although he continued to work on other cosmological models he never again used it. Strangely, though, others were reluctant to let it go, and it is still seriously considered in theories even today.

Incidentally, it was during the tour of the observatory that Einstein's second wife, Elsa, upon being told that the giant telescope was needed to determine the structure of the universe, replied, "Well, well, my husband does that on the back of an old envelope."

Just before Einstein left California he delivered a speech. Everyone expected a talk praising scientific progress, but Einstein surprised them by asking why scientific progress had brought so little happiness. He pointed out that it was used extensively for war, and even when there was no war it made man a slave to machines.

At almost the same time that Einstein was giving his speech, across the Atlantic in London, Eddington was addressing the Royal Astronomical Society. "It's surprising," he said, "why there are only two solutions to Einstein's equations." He went on to talk about the lack of time-dependent solutions—solutions that might correspond to Hubble's recent discoveries. His talk was published in the Society's *Monthly Notices* where it came to the attention of a former student of his, Georges Lemaître. Lemaître immediately wrote Eddington informing him that he had published a time-dependent solution in a Belgian journal. Eddington looked it up, and was so pleased with it he had it republished in the *Monthly Notices*.

LEMAÎTRE'S MODEL

Georges Lemaître was born in Charleroi, Belgium, in 1894. In 1911 he enrolled at the University of Louvain as an engineering student but his studies were interrupted by World War I. After serving in the army for four years he returned to the University of Louvain in 1918, but switched to mathematics and physics. Soon after graduation he entered a seminary and in 1923 he became a Catholic priest. In the same year he was awarded a government scholarship that allowed him to study

Georges Lemaître.

abroad. He used it to spend a year at Cambridge working under Eddington, and to study at Harvard and MIT in the United States.

While in the United States Lemaître attended a lecture by Hubble at the National Academy of Sciences meeting in Washington, D.C. Hubble talked about his work with nebulae, and his proof that they were island universes of stars, distant from the Milky Way. (He had not yet begun to work on redshifts.) The talk sparked Lemaître's interest in cosmology and soon after his return to Belgium he published his first, and most important, paper on cosmology. Examining Einstein's equations (with the cosmological constant) he found several time-dependent solutions that corresponded to an expanding universe. Out of them he singled one that seemed more relevant than the others. In it the universe began as a "big bang," expanded until it became a static Einstein-like universe, then became unstable again and ended as a de Sitter universe (see figure p. 51). Lemaître

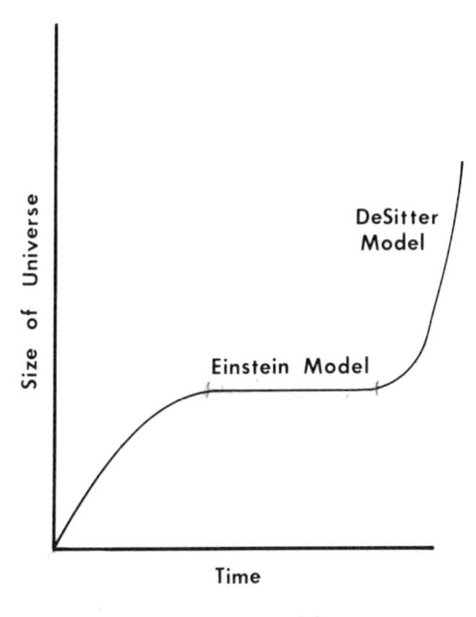

Lemaître's model.

believed that the galaxies may have formed during the Einstein static stage.

FRIEDMANN'S MODEL

But a time-dependent model had already been discovered. Aleksandr Friedmann, in Russia, found in 1922 that if he discarded the cosmological constant he got an expanding universe. Born to a musical family—his father and grandfather were both composers—in St. Petersburg (now Leningrad) in 1888, Friedmann attended St. Petersburg University, specializing in mathematics, and later in theoretical meteorology. In 1914 he volunteered for the war, eventually becoming the director of a factory that manufactured aviation instruments. After the war he returned to St. Petersburg University, where he taught mathema-

A. Friedmann.

tics and physics and conducted various experiments in meteorology.

With the verification of Einstein's general theory of relativity in 1919, Friedmann became excited about it. He was cut off from most other scientists, and therefore had to learn the theory on his own. But within a short time he had not only mastered it but was making contributions to it. He noticed that Einstein had made a mistake in his proof that the universe was stable: he had divided his equations by an expression that in certain cases went to zero. Making the appropriate adjustments Friedmann found that he got a universe that expanded. In fact, he got several alternative models within his theory.

He wrote a paper and sent it to Einstein for his approval. After a few months of hearing nothing he began to worry. Then, upon hearing that another professor from St. Petersburg was going to Berlin, he asked him if he would visit Einstein and ask about the paper. A few weeks later Friedmann got a "grumpy" letter from Einstein acknowledging the paper and agreeing that it was publishable. Friedmann then sent it to the German journal *Zeitschrift für Physik* and in 1922 it was published. He began his paper with the statement, "The purpose of this note is to show that the cylindrical and spherical world [of Einstein and de Sitter] are special cases of a more general assumption, and secondly to demonstrate the possibility of a world in which the curvature of space is independent of the three spatial coordinates but does depend on time. . . ."

The paper, although published in an important journal, drew little attention. The only attention, it seems, came from Einstein, and it was strange in that he had already informed Friedmann he thought the paper was publishable. He sent the editor a short note pointing out what he thought was an error. Friedmann saw Einstein's note and quickly checked his calculations, discovering that he had not made a mistake. But what should he do? He was reluctant to challenge Einstein in print, so instead of writing to the editor he wrote directly to Einstein pointing out the mistake. He closed his letter with, "In the case that you find my calculation to be correct . . . will you perhaps submit a correction?" This was brave on the part of Friedmann—but it worked. Einstein wrote another note that was published several months later. In it he admitted his error, stating, "I am convinced that Mr. Friedmann's results are both correct and clarifying."

It might seem strange that Einstein had so little interest in the paper. But Hubble had not yet advanced his expanding universe theory and Einstein was still convinced that it was static. He no doubt tended to think that it was merely an exercise in mathematics.

Eventually, though, he came to accept Friedmann's theory, devoting many pages in his book *The Meaning of Relativity* to it. In this book he referred to his disgust and difficulties with the cosmological constant. He stated, "The mathematician Friedmann found a way out of the dilemma. His results then found a

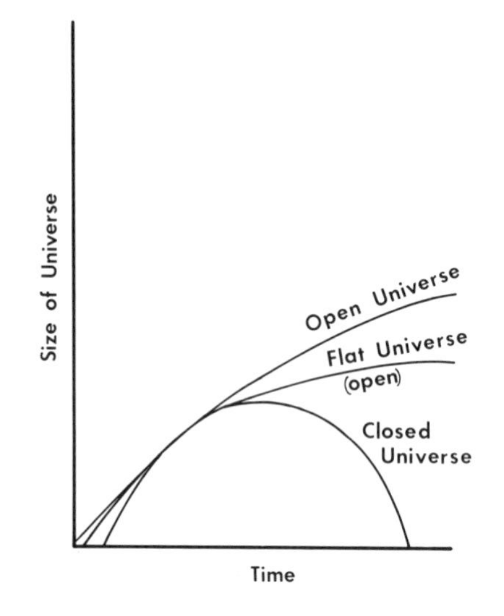

The three possibilities within Friedmann's model. If the universe is open, it will expand forever; if it is closed, it will collapse back on itself.

surprising confirmation by Hubble's discovery of the expansion of [the universe]."

Friedmann also published a second paper in 1924. The following year he died of typhus.

Let us take a moment to look at Friedmann's model. After discarding the cosmological constant he found that the equations gave three possible universes. They were: a positively curved, a negatively curved, and a flat one. The flat and negatively curved universes expanded forever, but the positively curved one collapsed back on itself. They can be represented simply as shown in the figure.

Which of these corresponds to our universe? To answer that we have to know the average density of matter in it. If it is over a certain critical amount the universe will be positively curved and will collapse; if not it will expand forever. In a sense the explosion that created our universe is like an ordinary explosion here on Earth. We know, for example, that if an explosion is powerful enough the pieces will be blown completely free of the Earth. In other words, they will overcome the gravitational attraction of Earth and move off into space. If they do not have enough energy, on the other hand, they will fall back to Earth. In Friedmann's theory, if the explosion was powerful enough, the matter of the universe will separate forever. Or, if the mutual gravitational attraction of the matter within it is sufficiently great, it will eventually collapse back on itself.

OTHER MODELS

Eddington was at first excited about Lemaître's cosmology but as time passed he became disenchanted with it. He also found fault with Friedmann's model. His main concern was with the "big bang"—the explosion that began everything. A universe that was suddenly "created" a few billion years ago was repulsive to him. He could not bring himself to think about such a model.

Lemaître, on the other hand, preferred a beginning, no doubt because of his religious beliefs. It meant that the universe had not been here forever, but had been created some time in the past. Much of Lemaître's later career was, in fact, centered on explaining this beginning. He visualized what he called a "primeval atom," a small dense state of nuclear matter that suddenly became unstable and exploded. He even went as far as trying to show that the elements we now have in the universe were generated in the explosion.

Because of Eddington's abhorrence of beginnings he developed a model that began in an Einstein state, became unstable, and then expanded into a de Sitter state. In his model the universe was in an Einstein state in the infinite past. It is hard to believe that Eddington took this seriously as he was the one who proved that Einstein's universe was unstable. A small perturbation and it would expand or contract. If this was the case how could it have stayed in such a delicately balanced position for an infinite amount of time?

Einstein took a last stab at cosmology in 1932 shortly after he abandoned his cosmological constant. Together with de Sitter he developed a model that did not contain the constant. It was a particularly simple model that expanded forever. But upon detailed examination it was shown to have unsatisfactory properties and was soon abandoned.

REFLECTION

So far we have said little about the creation of the universe. The reason, of course, is that we had to see how and why such a creation occurred. If the universe had an infinite past then there would have been no creation. But Hubble showed us that the universe is expanding, and coupling this with the cosmological models of Friedmann and Lemaître we see that the universe had to have a beginning. It had to have been created.

When was it created? This is, of course, the same thing as asking: What is the age of the universe? There is still considerable controversy over this, and we will discuss some of the problems later, but it is now generally agreed that it is about 16 to 18 billion years old. This means that the "big bang" explosion occurred at this time. And from this explosion arose the universe. Beginning in the next chapter and throughout the rest of the book we will be looking at the details of this explosion and the period immediately after it.

CHAPTER 4

Alpha, Beta, and Gamow

The universe was expanding, or at least there was considerable evidence it was expanding. With Hubble's observations and a prediction based on Einstein's general theory of relativity few were willing to argue that it was not. This meant the galaxies were moving away from us, and the farther they were away, the faster they were moving. But not only are they moving away from us, they are also moving away from one another. It is the space between the galaxies that is expanding. A closer look reveals, though, that many of the galaxies are in groups, or clusters, and the individual galaxies of these clusters are not moving away from one another—their mutual gravitational attraction is too great. Therefore, to be perfectly accurate, we should say that clusters of galaxies are moving away from all other clusters. But any way you look at it the universe is still expanding. And by the late 1930s it was clear that Friedmann's theory was the most acceptable of the several theories that explained this expansion.

Let us take a moment to consider this expansion. All galaxies are moving away from us, and each other. This means that in the past they were all closer together. In fact, the further we go into the past the closer they were together. If we go back far enough we therefore eventually get to a point where they were all in one small region—or at least their matter was in one small region of space. Lemaître referred to this as the "primeval atom."

THE PRIMEVAL ATOM

Could all of the matter of the universe actually be packed into a sphere about the size of the Earth's orbit? Admittedly, it is difficult to visualize, but this is the way Lemaître pictured things. He thought of his primeval atom as made up of protons and electrons. It is not clear whether he thought of it as in a finite or infinite space, but he did think of it as in some sort of empty space. Lemaître's ideas are the basis of the way we think of this "atom" today, but we have quite a different view. Lemaître thought of his atom as extremely dense; today we assume it was infinitely dense. Furthermore, we prefer to think of it as having no dimensions. Such an object is called a singularity. And finally, we prefer to think of the entire universe, space and all, as tied up in this singularity. In other words, it did not exist in an infinite empty space, it existed in "nothing." Don't worry about what "nothing" means for now; we will come back to it in a later chapter.

Anyway, getting back to Lemaître's model, we find that he was guided by a discovery that had been made several years earlier: radioactive decay. Antoine Becquerel showed in 1896 that atoms could spontaneously disintegrate. He noticed that uranium gave out strange "rays," and in the process, changed to a lighter element. Lemaître believed that his primeval atom decayed in a similar way. Today it seems more natural to compare his decay to what we call nuclear fission (a breaking apart or splitting of the atom). You are likely familiar with fission in relation to the atomic bomb; it is the process that causes the explosion.

Fission was, of course, unknown to Lemaître and when he first formulated his ideas even the atom was not fully understood. Scientists had accepted Rutherford's suggestion that it was composed of a heavy nucleus, surrounded by orbiting electrons. But the neutron, an uncharged particle we now know is in the nucleus, had not yet been discovered. So, needless to say, Lemaître's model was crude. One of its shortcomings is even evident in its name. Lemaître referred to it as the primeval atom, but what he had in mind was actually much more like a nucleus, and it should therefore have been called the primeval nucleus.

Lemaître spent several years developing his ideas. Once the neutron had been discovered he began thinking of his "atom" as composed of neutrons, rather than protons and electrons. His major contribution, it can be said, was that he gave the big bang theory a physical counterpart. Before Lemaître we had only a set of equations based on general relativity that told us how the universe expanded. But they told us little, or nothing, about the beginning of the universe. Lemaître supplied the details, and for this reason he is now frequently referred to as the father of the big bang, a title he certainly deserves.

Lemaître was also interested in predicting the abundance of the elements in the universe. He was, in fact, convinced his theory would allow him to do this. But it was a question that was ahead of its time. The abundance of the elements had not yet even been measured.

For the most part Lemaître's description of the fragmentation of his atom was nonmathematical. Furthermore, he did not follow up on the details. This was left to Maria Meyer and Edward Teller of the University of Chicago. They assumed a "superatom" about 15 miles across composed of a nuclear fluid, and showed that bumps, or "pimples," would form on its surface soon after it formed. These pimples would then explode, throwing off matter into the space around the superatom. But difficulties eventually arose and they had to abandon their model.

THE ABUNDANCE OF THE ELEMENTS

Before we look at the developments that came shortly after Meyer and Teller's work, let us go back to the problem of the abundance of the elements. Looking out into the universe we see a curious distribution. The simplest and by far the most abundant atom is hydrogen. By weight, over three-quarters of the matter of the universe is hydrogen. This is perhaps reasonable in that the hydrogen atom is particularly simple, made up of a nucleus of a proton, with a single electron in orbit around it. The electron is held in orbit by the electrostatic force between the opposite charge of the proton and the electron.

Aside from hydrogen, most of the rest of the matter of the universe is helium. This may seem strange as helium is quite rare here on Earth. So rare, in fact, that it was discovered in the sun before it was discovered on Earth. Its nucleus is made up of two neutrons and two protons, with two electrons in orbit around it. The bare nucleus, in other words the cluster of two protons and two neutrons, is called an alpha particle. There are nuclei that are simpler than helium-4 (the 4 here refers to the number of particles in the nucleus), for example, deuterium and tritium, which are heavy forms of hydrogen, and there is a lighter form of helium that is called helium-3. Yet helium-4 is much more common than them; almost one-quarter of the matter of the universe is made up of it. Why? It's important that we be able to answer this. And we will—later.

Most of the rest of the elements make up only about 1% of the total, and of these the most common by far are carbon, oxygen, and iron. Carbon has a nucleus made up of 6 neutrons and 6 protons. Iron, on the other hand, has 26 protons and 30 neutrons in its nucleus. Incidentally, most of the weight of the atom is in the nucleus, for the proton and neutron are two thousand times as heavy as the electron. Each element is, in fact, characterized by its atomic weight, which to a first approximation is how much heavier it is than hydrogen (more exactly, it is based on oxygen being 16).

In the 1860s the Russian, Dmitri Mendeleev, listed the atomic weights of all the elements on cards and began grouping those with similar chemical properties in columns. And lo and behold he found he could form several such columns; lithium, beryllium, and carbon, for example, fell into one column. Sodium, magnesium, and aluminum fell in another. And so on. The result of his "card playing" was what we refer to as the periodic table. There are now over 100 elements divided into eight columns in this table.

As I mentioned earlier, each of the elements of the periodic table has a certain abundance in the universe, an abundance we have now measured. How do we measure it? We can easily determine the abundance in the Earth's crust by checking it directly. But we also have access to the stars and the matter between the stars. By studying their spectra we are able to determine what elements they contain. And surprisingly, the distribution is quite uniform throughout the universe.

Earlier I said that the most common elements are hydrogen and helium. But they are certainly not the most common on Earth, and there is a reason for that. If we go back to the time shortly after the Earth was formed we find that it had a lot of hydrogen and helium—it had a dense atmosphere composed of these elements. At that time all the inner planets were covered by an atmosphere of this type, much as Jupiter and Saturn are today. But when nuclear reactions were triggered in the sun a shock wave moved out from its surface—a solar gale—so powerful that it blew the atmospheres from the inner planets, and left only the heavier elements that had accumulated at the core. This gale was, incidentally, not strong enough to blow the atmospheres from Jupiter and Saturn, and that is why they still have them today.

Now, let us take a look at this abundance curve (see figure). The first thing we notice is that, to a first approximation, the heavier an element, the less of it there is in the universe. There are, of course, anomalies—iron being the most glaring one. But there is also something else that is strange; when we get to elements above atomic weight 100 or so, the abundances are about equal. In other words, the curve levels off for the heavy elements. And there are other enigmas. For example: why are hydrogen and helium so common? And why is there so little lithium and beryllium? And why is iron much more common than the elements around it? All of these questions can be summed up in: Why does the curve have the shape it does? If

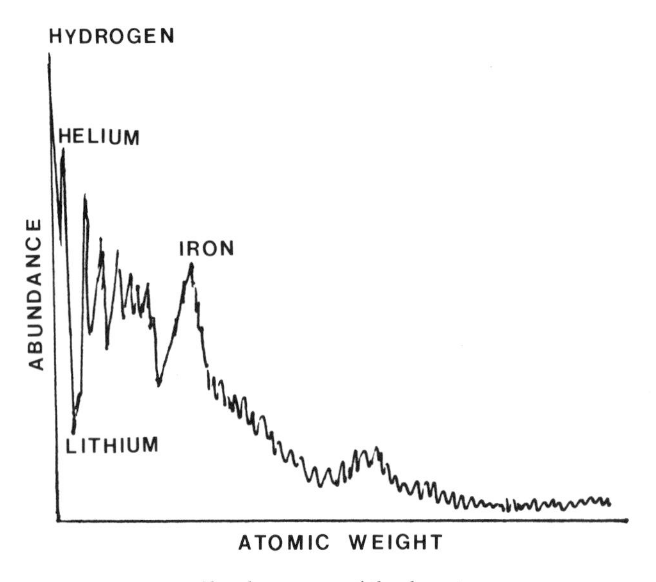

Abundance curve of the elements.

you think about it for a moment you realize that an important key to the understanding of creation is an understanding of how this curve came about. But to understand it we have to know where and how the elements were generated. One possibility, theorists realized early on, was in the stars. But a snag soon developed and other possibilities had to be considered. The only alternative seemed to be the big bang explosion, and for a while there was considerable interest in it. Then it too developed a snag. In the end, though, everything worked out and we now believe that both stars and the big bang make important contributions.

Okay, now let's go to the details. First, we have to assume we have some initial substance. Assume that it is hydrogen. How, then, do we get heavier elements from this hydrogen? The first clue came when scientists discovered that if they projected a neutron at it with enough energy, the neutron would "stick" to it and we would get deuterium. Similarly, if we project another neutron at the deuterium nucleus, we get a tritium nucleus. Could all of the elements be built up in this way? Let us look at this.

EARLY EXPLANATIONS

One of the first to try to explain how the elements formed from one another was Carl Von Weizacker. He assumed that the buildup occurred in the interior of stars through a series of neutron bombardments and decays. Deep inside stars temperatures are exceedingly high, and neutrons easily have enough energy to "stick" to nuclei when they strike them. One thing I did not mention earlier, though, is that not all nuclei are stable. The tritium nucleus, for example, is one that is not; soon after it is formed one of its neutrons spits out an electron and becomes a proton. The process is called beta decay. Before we continue, let us take a moment to consider it.

The neutron is a stable particle as long as it is in an atom. The neutrons in the helium nucleus, for example, remain neutrons indefinitely, as long as they are part of the nucleus. If, however, you pull one of them out so that it is free, then in about thirteen minutes it will decay. In other words, it will change into a proton and an electron. Actually, another particle is also released, but we will ignore it for now. In some cases, though, the neutron need not be free to decay. The tritium nucleus is a case in point.

Weizacker's work was picked up and extended by Hans Bethe. Born in Germany in 1906, Bethe attended the University of Frankfurt and Munich, obtaining his Ph.D. in 1928. With Hitler's rise to power he departed Germany in 1933 and went to Britain, then a year later to the United States. Bethe considered nuclear reactions in the light elements, showing how helium-4 could be produced from hydrogen. He hoped to show that all of the elements could be built up by collisions and decays but was stopped by element number 4. Helium-4 was so stable that when a neutron struck it, although it briefly gave an element with 3 neutrons and 2 protons, this new element quickly decayed. In short, there was no stable element number 5—just a gap at position 5. There was the possibility of jumping over this gap, but Bethe could not see how it could be done. He wrote, ". . . under present conditions, no elements heavier than helium can be built up [in stars] to any appreciable extent."

S. Chandrasekhar and L. R. Henrich also worked on the problem. In 1942 they wrote, "It is now generally agreed that the chemical elements cannot be synthesized under conditions now believed to exist in stellar interiors." Not only was element 5 a problem for the early workers but they could also not explain why the abundance curve leveled off about halfway through the elements.

A way around the difficulties was soon recognized by George Gamow. Or at least he thought it was a way around. So far everyone had been considering stars. Gamow turned to the early universe, and became convinced that the elements were formed in the big bang. But the universe cooled off rapidly and the process could not possibly be an equilibrium one. He therefore turned to nonequilibrium processes.

GAMOW

George Gamow loved life, loved a challenge—and reveled in practical jokes. Born in Russia in 1904, he lost his mother early in life and was raised by his father. His father's passion was opera; he lived opera, sang opera, and hoped to instill a similar passion in his son. But to no avail. Gamow did, however, take a particular interest in one opera: *Russlan and Ludmilla*. It was the story of a giant who was beheaded. Gamow was eager to see how they were going to perform such a feat on stage. But when he went to the opera, much to his dismay, it occurred out of his sight.

George Gamow.

Russia was at war during most of Gamow's youth, and his education was frequently disrupted. He was standing by a large window one day when a nearby exploding shell shattered it. But this and other incidents did not deter him from developing an intense interest in astronomy and physics. His interest in astronomy was heightened, in fact, when his father gave him a telescope for his 14th birthday. He studied at Novorassia University and the University of Leningrad, graduating in 1928, about the time quantum theory was blossoming. After receiving his Ph.D. he was lucky enough to spend several years at three of the institutions that were at the center of quantum mechanical development: the University of Göttingen, the Niels Bohr Institute at Copenhagen, and Cambridge University in England. It was an exciting time in physics. Cafes and seminars were crowded with physicists arguing about the consequences of quantum mechanics. But as Gamow said, ". . . somehow I was not engulfed in this whirlpool of feverish activity. One reason was that too many people were involved in it." He preferred to work in a less crowded area. Realizing that most of the activity centered around the application of the new theory to atoms and molecules, Gamow decided to apply it to the nucleus.

One of the major problems of the day was the spontaneous decay of the nucleus—referred to as alpha decay. The uranium nucleus, for example, emitted alpha particles, but there was something strange about the process. If an alpha particle was projected at the nucleus it was deflected; there was a strong barrier preventing it from getting in. How could alpha particles come out through this barrier if it was impossible to penetrate it from the outside? Gamow used quantum theory to show that the barrier was penetrated in an odd way: the alpha particle "tunneled" through it. This was the first successful application of quantum theory to the nucleus, and it was also one of Gamow's major contributions to physics.

Gamow was described by all who knew him as a fun-loving man of boundless energy. Max Delbrück, who roomed with him at Copenhagen, described him as "very tall and thin, looking even taller for his erect carriage, blond, a huge skull, and a grating high-pitched voice." Delbrück went on to say, "I would go to bed around eleven, then about midnight Gamow would come in, turn on the lights, unpack beer and hot dogs and discuss the evening's adventures: what she had said, and what he had said, or what practical joke to play tomorrow."

Practical jokes were Gamow's forte. Within days of meeting him, said Delbrück, "Gamow poured liquid air into my prized black bowler hat and dropped it on the floor. When a large piece broke out of the crown he sent it as a postcard to a friend in Göttingen." It was about that time, however, that Gamow was thoroughly outdone in the practical joke department. Three postdoctoral fellows at Cambridge, G. Beck, H. Bethe, and W. Reizler, wrote a spoof (a meaningless paper that sounded impressive) in the style of Eddington, and sent it to the journal *Naturwissenschaften*. After it was published the editor found out it was a practical joke. He was outraged and demanded a written apology.

Gamow thought it was hilarious—but he had been outdone. He would need a comeback to preserve his reputation. He waited for another similar, but legitimate paper, then convinced Wolfgang Pauli and L. Rosenfield to write a letter to the editor telling him they were annoyed that he had fallen for another practical joke. With three letters coming in (one also from Gamow), the editor was confused, and annoyed. As it turned out, though, they did not convince him.

Aside from pulling practical jokes Gamow was a poor speller, had trouble with addition and subtraction, could not remember names or faces, but he did have a flair for poetry—limericks in particular (some of them unprintable). He once bet someone he could continue, nonstop, quoting verse for one hour. The bet was taken up—and Gamow won. He continued for an hour and a half, and quit only because everyone got bored.

In 1933 Gamow came to the United States and soon had a position at George Washington University in Washington, D.C. In later years he began writing popular books and eventually became well known as a popularizer.

While at George Washington University he was a consultant to the Applied Physics Lab of Johns Hopkins. Two students, Ralph Alpher and Robert Herman, also worked at Johns Hopkins, and Gamow eventually met them. Alpher became interested in Gamow's work and began working on a master's thesis under him. Upon completion of it he turned to a Ph.D. thesis, also under Gamow, on turbulence in galaxies. But he had barely got started when Lifshitz, a Russian, scooped him by publishing a comprehensive paper on the subject. Alpher was therefore forced to look around for a new thesis topic.

His problem was solved in January of 1946. The program of the New York meeting of the American Physical Society described some work by D. Hughs of Brookhaven National Laboratory. Hughs had measured the cross sections of several of the light elements. (Cross section is a measure of the probability that a certain type of nuclear reaction will occur. It is a number that tells you, for example, if helium-4 is likely to capture a neutron that is shot at it.) Gamow realized that with these cross sections

Ralph Alpher (left) and Robert Herman.

the problem of element abundance could be tackled again. As we saw earlier it had been dropped when scientists realized—or thought they realized—that elements could not be produced in stars. Gamow now looked to the early universe. Perhaps the elements were produced in the big bang explosion. It was important, though, Gamow pointed out to consider nonequilibrium processes.

So Alpher had a new problem: the nonequilibrium formation of the elements by neutron capture in the early universe.

ALPHA, BETA, GAMMA THEORY

The first thing to determine was what the initial material in the universe was like. Gamow, however, already had ideas on this. He had even given it a name—ylem (which, oddly enough, means first material). Where Lemaître had assumed a nucleus that broke down into lighter and lighter particles, Gamow visualized a building-up process that started with a dense, yet gaseous nucleus consisting mostly of neutrons. But as we saw earlier, free neutrons beta decay to give protons and electrons, so protons and electrons would also be present. Under the hot, dense conditions in Gamow's nucleus, though, it turns out that we would also have the same reaction in reverse. In other words the protons would absorb electrons and become neutrons. So for a while we would have both reactions going, and the number of particles (of a particular type) would remain relatively constant.

But the universe was rapidly cooling at this stage and soon it was too cool for protons and electrons to become neutrons, and we were left with only decaying neutrons. Of course, some of the protons would be hit by neutrons and become deuterium nuclei, or deuterons. Similarly deuterons would be hit, producing heavier nuclei, and so. To guide him, Alpher now had experimental values for their cross sections. He began by drawing a smooth curve through the points representing the cross sections. The scatter in the points, particularly for the very light elements, was large, but as we saw earlier the points on Hubble's redshift–velocity graph also had a large amount of scatter, and he drew a smooth curve through them. It seemed reasonable to do the same thing here.

Everything progressed well and Alpher wrote up his thesis. Periodically he would meet with Gamow, giving him a summary of his progress. True to his independent spirit Gamow insisted that their meetings take place, not at the university, but at a bar and grill called "Little Vienna," near the campus. During these sessions there was lots of gossip mixed with a few drinks, but of course considerable work was also accomplished.

It was at one of these meetings. when Alpher was nearing completion of his calculations, that Gamow got a gleam in his eye. He had not played any practical jokes lately. It was about time. He pointed out that Alpher's name and his sounded like the first and third letters of the Greek alphabet, alpha (α) and gamma (γ). But he needed a beta (β) to complete the sequence. Then it came to him—Hans Bethe of Cornell. They would publish the paper in *Physical Review*, calling it the $\alpha\beta\gamma$ theory. Of course, as Bethe did not contribute anything, they would add "(in absentia)" after his name. Alpher, and Herman, who was now also involved with the group, were not strongly in favor of the plan, but Gamow insisted, and they went through with it. If Bethe objected, Gamow said he would remind him of his earlier spoof of Eddington.

Much to Gamow's delight the paper was published on April Fools' Day, 1948. And Bethe did not object. In fact, he thought it was quite humorous, and later came to Alpher's thesis defense. The strange part is that the "(in absentia)" somehow got erased in the final editing and the paper appeared without it. Gamow was pleased with the publicity that the paper generated. Several feature articles were published describing it. It got so much publicity, in fact, that Alpher's thesis defense became a public event. Over 200 people attended, including several from the press.

But finally it was over, and time for celebration. Gamow supplied a bottle of well-known liqueur on which he had replaced the label with the name "ylem." After polishing off the bottle Alpher and Herman took it home. Now it was time to play a joke on Gamow. They made up a photographic montage using the bottle with its ylem label, showing Gamow coming out of the bottle as a genie. Then, without him knowing, they slipped it into the set of slides he used in his public talks. When he gave his next talk he was surprised—but loved it, and frequently used it after that.

PROBLEMS

But all was not well with $\alpha\beta\gamma$ theory. During a presentation of the theory Enrico Fermi was in the audience. He was already

Gamow as genie.

famous as the producer of the first sustained nuclear reaction, a process that eventually led to the atomic bomb. He was not comfortable with the fact that Alpher had drawn a smooth curve through the points representing the cross sections—particularly in the case of the light elements where there was a lot of scatter. The neutron capture cross section of helium-4, for example, was almost zero, whereas nearby nuclei had relatively high cross sections. Fermi had his student, A. Turkelvich, redo the calculations using the exact cross sections (not averages) and found what Gamow was already becoming suspicious of: there was a gap where element number 5 should have been, just as there had been with stars. The only way, it seemed, that this gap could be overcome was by the simultaneous capture of two neutrons. If this happened we would get lithium, but the probability that it did was exceedingly small. Furthermore, even if they could get around the gap at 5, there was another gap at 8. Things looked bleak.

Fermi, however, took things further. If the elements were not produced in the early universe, maybe we should take another look at stars. Bethe had had problems with stellar synthesis but a lot had been learned since 1939. Fermi asked Martin Schwarzschild to look for evidence of heavy elements in stars. Schwarzschild and his wife made a thorough study of faint stars. From their spectra they found that there was ample evidence of heavy elements. They found that the large blue stars had a greater abundance of iron and other heavy metals than the smaller red ones did. There was now no doubt: heavy elements were, indeed, formed in stars. But there was still the problem that Bethe had encountered (and also Gamow in relation to the early universe): how to get past helium-4.

Schwarzschild presented the problem to one of his students, Ed Salpeter. Armed with much better experimental information than Bethe had been years earlier, Salpeter soon found a way around the problem. And interestingly it was a solution that had been briefly considered, but discarded, by Bethe. The gap could be hurdled because beryllium-8, although unstable, provided a stepping-stone for a jump to carbon-12. Thus the element that was formed after helium in stars was carbon, a seemingly large jump, but evidently not an impossible one.

The early universe was no longer needed for the production of elements—or was it? As it turned out it was. Fred Hoyle of Cambridge was examining the energy released by galaxies as a result of their helium formation when he discovered it was ten times too high. There was no way all that helium could be generated in stars—there was just too much. So back to the drawing boards they had to go. The only other place helium could be

Jumping the gap at 5.

produced was in the early universe. And indeed, Gamow and others had shown that there was no problem with helium—it was the elements above helium. On the basis of this we now believe that most of the helium and other light elements were formed in the big bang whereas all the other elements were formed in stars.

LATER STUDIES

Although few were interested in the early universe once it was established that the elements could be formed in stars, Alpher and Herman continued working in the area. They were later joined by James Follin. They felt that it was important to understand what had happened immediately after the big bang. Previously they had dealt mainly with neutrons, protons, and electrons. But detailed study showed that there were also other particles present. Photons, or "particles" of radiation (light), would also be present in large numbers. Furthermore, when I talked about beta decay earlier I mentioned that another particle was released, besides the proton and electron. We now call this other particle a neutrino. It was also present in the early universe in large numbers, along with the "antiparticle" to the electron—the positron. I will talk about all these particles in more detail later.

All these other particles had to be taken into consideration. And they were in a paper that was published in 1953 by Alpher, Herman, and Follin. It was titled, "Physical Conditions in the Initial Stages of the Expanding Universe." Starting at about 10^{-4} second after the big bang they found that within a short time the universe was dominated by photons. In other words, there was a lot more radiation around than matter particles. This remained true for about 10,000 years. They calculated the ratio between proton numbers and neutron numbers and noted how it changed in time as the universe expanded and cooled. And they also followed the history of the neutrinos, noting that although they were initially in equilibrium with other particles they eventually "froze out" and moved off freely into space.

Their work was followed up by Peebles and by Fowler, Hoyle, and Wagoner several years later, who made more detailed calculations of the abundances of the light elements in the early universe.

CHAPTER 5

From Quarks to Black Holes

With the work of Gamow, his students, and others it became obvious that particle physics was the key to an understanding of the early universe. Because of this, particle physicists soon began to take an interest in cosmology. And gradually a symbiosis of the two fields began which turned out to be beneficial to both. The cosmology of the early universe allowed particle physicists access to the greatest particle accelerator ever built—the big bang explosion. Incredible energies, infinitely higher than anything we can create here on Earth, occurred in the first few moments of this explosion. And the mathematical description (theory) of the explosion puts severe constraints on the types of particles that can exist in the universe. This has been particularly helpful to particle physicists. The benefit of the union to cosmologists has been equally great. With particle physicists joining in the struggle much of the arbitrariness and speculation has been taken out of cosmology; it has been made into a more testable, more respectable science. At one time you could get away with almost any type of speculation. No more. The merging of particle physics into cosmology has changed it forever.

To understand the details of the early universe we must begin by understanding the particles that inhabited it, and the forces between them.

PARTICLES

What is the universe made of? One way to find out is to take a microscope and look at the materials around you. Unfortunately, even with the largest microscope we cannot see atoms, at one time considered to be the fundamental building blocks of matter. Yet we have considerable evidence that they exist. In fact we now know that they are made up of protons, neutrons, and electrons. Are these particles, then, the truly fundamental particles of the universe? To answer this we would need a bigger and better "microscope," an instrument that could somehow penetrate these particles. And, indeed, we have one: the particle accelerator. With particle accelerators physicists are able to probe deep inside protons and neutrons.

A particle physicist is, in many ways, like a small boy with a shiny new marble wondering what is inside of it. How does he find out? He takes a hammer and smashes it. Particle physicists do the same thing. To find out what was inside the proton they projected electrons at it. And—alas—they discovered that protons are not fundamental particles; they are made of even simpler particles—quarks. Bouncing around inside the proton are three of these quarks.

But even before the first experiments had been performed showing the existence of quarks, they had been predicted by Murray Gell-Mann (now of Caltech) and independently by George Zweig (then of CERN). The name "quarks" was, in fact, coined by Gell-Mann; Zweig preferred to call them "aces," but it was the name "quarks" that stuck.

Large accelerators have not only simplified things by showing us that many particles are made of quarks; they have also complicated things. As particle after particle was bombarded, new particles came out in the collisions. It was like shooting the cue ball at, say, the 8 ball and having the 1, 3, 5, and 7 balls coming out of it. The new particles were given names such as pions, kaons, and strange particles. But most of these particles were not fundamental; like the proton and neutron, they too

Tevatron at Fermilab. (Courtesy Fermilab.)

were made up of quarks. A few of them, however, were not. The electron, for example, was not, and neither were two close cousins to it—similar but heavier particles called the muon and tau.

We now know that all the matter particles of the universe are composed either of quarks, or are similar to the electron (or related to it). There are, in fact, two classes of particles called quarks and leptons.

Let us consider the quarks first. They come in pairs, referred to as up and down, strange and charmed, and bottom and top. Using shorthand notation we refer to them as

$$\begin{pmatrix} d \\ u \end{pmatrix} \quad \begin{pmatrix} c \\ s \end{pmatrix} \quad \begin{pmatrix} t \\ b \end{pmatrix}$$

Particle tracks in a bubble chamber. (Courtesy Fermilab.)

The proton, as I mentioned earlier, is composed of three quarks: two u's and a d. The neutron, on the other hand, is composed of two d's and a u. You can think of them as marbles bouncing around inside a plastic bag. But the strange thing is that they seem to be confined to the bag. As far as we know there is no way scientists can ever pull one from the bag. Free quarks do not seem to exist. If they try they just get other particles, which, in turn, are made up of quarks.

Earlier I talked about a particle called the pion. It is also composed of quarks, but it is different from the proton. It is not composed of three quarks; rather, it is composed of a u quark and a d antiquark. What is an antiquark? The best way to explain it is to consider a simple experiment. Think of the collision of a u quark and its antiquark (written $\bar{u}$). What would happen? It turns out that they would annihilate one another—both would disappear and in their place we would get photons. There are, it turns out, antiparticles corresponding to all particle types in the universe, and when any of these particles and their antiparticles are brought together they annihilate one another. Note, however, that a quark does not annihilate with a quark of a different type (e.g., u and $\bar{d}$).

Now back to the pion. It is one of a whole class of medium-weight particles called mesons, each of which is composed of a quark and an antiquark. The heavier particles such as the proton and neutron, on the other hand, are called baryons. They are all composed of three quarks, and their antiparticles of three antiquarks. Together, the baryons and mesons are referred to as hadrons.

Incidentally, there is something else about these quarks I should mention: they are colored. Red, blue, and green are the three colors usually assigned to them. I hasten to add, though, that the word "color" as it is used here has nothing to do with the usual meaning of the word. Neither, for that matter, do the terms red, blue, and green. Why, then, do scientists use them? More than anything, I suppose, it's just a pun. Even scientists

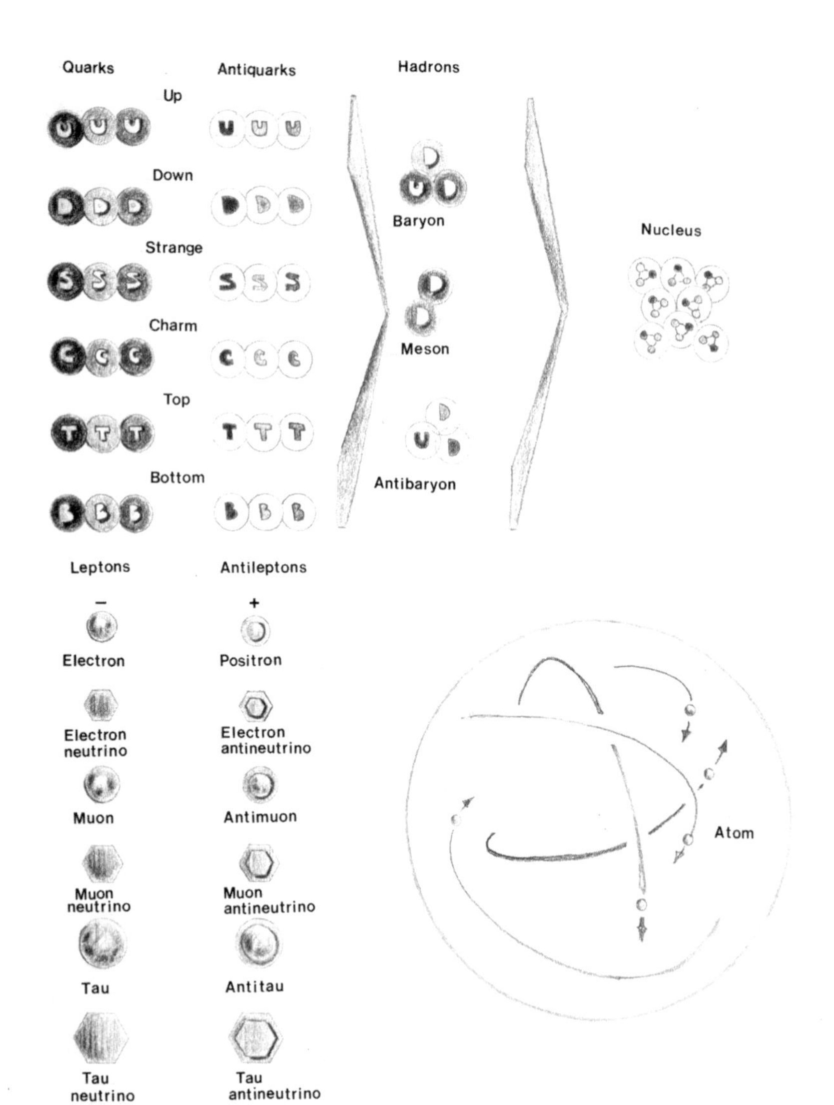

Elementary particles.

like to have fun once in a while. Color, in the microworld, is a force, like gravitation. It's what holds the quarks together.

Now to the leptons. As I mentioned earlier the electron is the most familiar one, but it has two cousins, the muon (μ) and the tau (τ). Both are heavier, the tau being the heaviest. To make things more complicated nature has associated with each of these leptons a particle called a neutrino. There is an electron neutrino (ν_e) a muon neutrino (ν_μ), and a tau neutrino (ν_τ). This means, then, that just as quarks come in pairs, so too do leptons. We can, in fact, write them in the same way.

$$\begin{pmatrix} \nu_e \\ e \end{pmatrix} \quad \begin{pmatrix} \nu_\mu \\ \mu \end{pmatrix} \quad \begin{pmatrix} \nu_\tau \\ \tau \end{pmatrix}$$

This may seem a little odd: a perfect matchup between quarks and leptons. To physicists, though, it is almost miraculous, something that has deep meaning. But something else is just as strange. All ordinary matter of the universe is made up of only the first bracket of quarks $\begin{pmatrix} d \\ u \end{pmatrix}$ and the first bracket of leptons $\begin{pmatrix} \nu_\mu \\ e \end{pmatrix}$—what we usually refer to as first-generation particles. The others are only created in huge accelerators. Of course, the early universe was a giant accelerator, and they were also created there. But now they do not seem to be needed. As to why they exist, we are not sure. It's as if we sat down to build a model of a wooden sailing ship and stocked up on iron, steel, and glass—just in case.

Something else about particles that is not well understood is why they are unstable. If left to themselves long enough they decay to lighter particles. An example is the neutron. If pulled from an atom, a neutron will decay in about 12 minutes to a proton, an electron, and an antineutrino. Only when we get to the simplest particles is there stability. The electron, as far as we know, is stable. And for years it was also thought that the proton was. But now we are not so sure.

Also: where do particles come from? Is it possible to make them, in the way we make a watch, or a TV set? Oddly enough, it is. All you need are the energy particles I talked about earlier—photons. The recipe goes as follows: bring the photons together at high speed and thoroughly mix. With the proper energy you can get almost any particle you want, along with its antiparticle.

Since this is particularly important in relation to the early universe let us look at it a little closer. Because we need energetic photons we will begin by talking about energy. It is usually measured in units called electron volts (eV). One way of visualizing an electron volt is to think of an electron moving between the positive and negative terminals of a flashlight battery—this is roughly 1 eV, a relatively small amount of energy. The particles in modern accelerators have energies of millions of electron volts (MeV) and even billions (GeV). Now, assume we have two photons, each with about 0.5 MeV, and we bring them together in a collision. What happens? We get an electron and its antiparticle (called a positron). If we up the energy to 135 MeV we get a pion and its antiparticle. And if we up it to 938 MeV, we get a proton and its antiparticle. And so on. We can, in fact, create any particle we want in this way.

This experiment is a direct verification of what Einstein told us many years ago: mass and energy are equivalent. Because of this equivalence we usually give the mass (or weight) of a particle in terms of the energy required to produce it. An electron, therefore, has a mass of 0.51 MeV, a proton, 938 MeV.

FORCES AND UNIFICATION

Closely associated with the two classes of fundamental particles of the universe are the four fundamental forces. You are no doubt familiar with at least two of the four. Gravity is the most common. It's the glue that keeps you stuck to the Earth. Without it you would float off into space. In fact, there would not even be an Earth, for it is held together by gravity. Gravity is different from the other forces in that it affects everything that has mass. We know of no way that it can be shielded—and it's a good thing (I would hate to think of the weapons that might result). As powerful as it seems, though, gravity is actually extremely weak—by far the weakest of the four forces. So weak, in fact, that atoms hardly feel it.

The second of our four forces is one you are also likely familiar with. It's electrical in nature, and referred to as the electromagnetic force. It's the force that holds atoms together, or perhaps I should say, holds the electrons to the nucleus. It is a trillion trillion times stronger than the gravitational field but we hardly notice it because it exists only between charges. The negative electron, for example, is attracted to the positive nucleus in the atom.

Both the gravitational and electromagnetic force are long-ranged. They act over long distances—infinite distances, in fact. The other two forces, as we shall see, are short-ranged. The first of these is the strong nuclear force; it is the glue that holds the protons and neutrons in the nucleus. It acts only over a distance equal to the size of the nucleus. But it is not the strong force itself that is fundamental. Just as the covalent force that holds molecules together is only a residue of the electromagnetic force, so too is the strong nuclear force a residue of the color force we talked about earlier. (The color force holds quarks together, but there's a little force left over that also holds the protons and neutrons together.) It is the color force that is the true fundamental force.

Finally we have the most difficult of the four forces to explain—the weak nuclear force. Like the strong force it is also short-ranged. One way of describing it is to say it is responsible for certain types of radioactive decay. You are likely most familiar with such decay in relation to radium, or uranium. They decay to lighter elements.

So much for the forces themselves. The next question is: How do they work? What I should say, perhaps, is: How do they "project" their force? Before quantum theory was invented we used to think of them as fields—an action-at-a-distance concept that was relatively easy to visualize but somewhat difficult to explain. An electric field, for example, was considered to be a kind of tension in the space around a charged particle—like a cobweb—that attracted other particles.

But with the coming of quantum theory the force concept took on a new meaning. The force field became a particle. When two electrons, for example, came near one another, they exchanged particles—photons—and were repelled from one another. It would be like two roller skaters throwing baseballs back and forth. As one catches the ball she is forced backward.

Because there are four different forces there must also be four different kinds of exchange particles. And indeed there are. Besides the photon we have the gluon (the exchange particle of the color force), the W particle (the exchange particle of the weak force), and the graviton (the exchange force of gravity). But if the forces of nature are actually particles we have another classification of particle: one associated with matter and one associated with force. We call these two types of particles fermions and bosons respectively. One way of distinguishing them is through their spin. Yes, you can think of particles as spinning too (actually, this is an oversimplified picture, but it will do for now). The electron, for example, is said to have a spin of 1/2. Other particles have spins 1, 3/2, and so on. Fermions, or matter particles, all have half-integral spin (i.e., 1/2, 3/2, . . .) whereas bosons, or force particles, all have integral spin (0, 1, 2,. . .).

Bosons are sometimes referred to as "connector" particles, because they "connect" the fermions of the world. But connectors do more than just bind fermions; they also change one kind of fermion into another. A W particle, for example, can change a muon into an electron. This makes us wonder if it is also possible to change leptons into quarks, and vice versa. In fact, because there is a symmetry, not only between quarks and leptons, but all through nature we wonder if we cannot do more than this: perhaps unify all particles (and all forces).

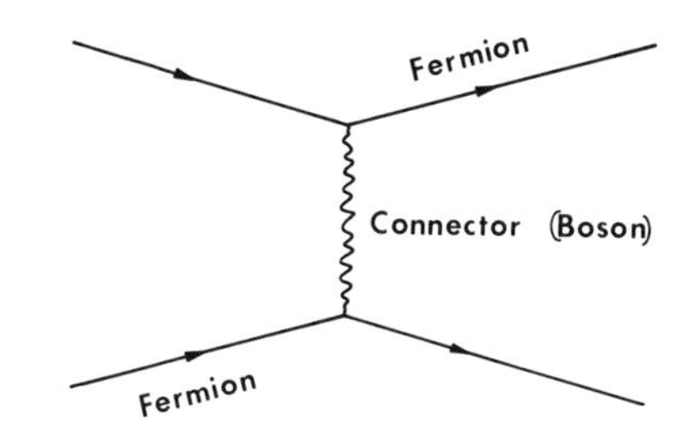

Simple diagram depicting a fermion–fermion collision.

Einstein was one of the first to try to unify the forces of nature, but when he began only two were known (gravity and electromagnetism). And, as you likely know, he was unsuccessful. Since his time, things have become more complicated: we now have four known forces of nature. And to unify them is a more formidable task. What do I mean by unify? Certainly we cannot make them all the same—we can easily distinguish, say, gravity and electromagnetism. What we would like to do is to have them be different "forms" of the same force—unified only under very different conditions than now exist on Earth. And, indeed, scientists have at least partially achieved this. They believe that at extremely high energies, those that existed in the very early universe, all four forces of nature were one—they were unified. But as the universe expanded and cooled one by one each of the forces we know today broke away from this unification.

We were led to this view by a breakthrough that occurred in 1967. Steven Weinberg of MIT and, independently, Abdus Salam of Imperial College in London showed that the electromagnetic and weak nuclear forces could be brought together, or unified, at energies greater than about 100 GeV. We now refer to this unification as the electroweak force. But how could these forces possibly be unified? Their exchange particles, the photon and W particles, are so different. The photon has no mass and is long-ranged, whereas the W particle has mass and is short-ranged. According to Weinberg and Salam the two particles are of the same family. The W particle has just "eaten" another particle called a Higgs particle, and in so doing has gained weight, whereas the photon has remained massless.

Spurred on by the success of the electroweak theory, two physicists at Harvard University, Howard Georgi and Sheldon Glashow, went a step further. They devised a theory in which the color force was unified with the electroweak force. We now refer to such theories as grand unified theories (GUTs, for short). In this theory there was a new exchange particle, a particle that changes a quark into a lepton (and vice versa), called an X particle. Compared to the exchange particle of the electroweak force (the W), though, it is incredibly heavy. It has a mass of 10^{15} GeV, which is far beyond anything we can achieve with present-day accelerators. In fact, we will never be able to build accelerators this powerful; they would extend beyond the stars.

Since the original GUT several others have sprung up, and we are still not sure which of them, if any, is correct. We will have more to say on this later. One thing I should say before leaving the topic, though, is that if the X particle exists the proton will no longer be stable, but will decay after an extremely long time—approximately 10^{31} years. Experiments are now in progress to see if this is the case.

BLACK HOLES

We are now in a position to talk about the early universe—almost. The reason I say almost is because besides particles in the early universe there are also objects we refer to as black holes. So to complete things we must talk about them.

The idea that black holes might exist came from the British clergyman and geologist John Michell. Interestingly, this was not Michell's only contribution to science; he also invented the torsion balance and created the branch of science known as seismology. About 1783 he began asking himself what would happen if gravity became so intense it would not allow anything to escape—even light. He knew that light traveled at 186,000 miles a second, and he was familiar with what we call escape velocity. This is the velocity needed to completely escape a given gravitational pull. (In the case of the Earth it is about 25,000 miles an hour. This means that if you blast off from the Earth at this speed you can completely escape its gravitational field, and will therefore not go into orbit around it.) Michell also knew that the more massive an object, the greater its escape velocity. This led him to wonder how massive a star (or any object) would have to be to have an escape velocity greater than the velocity of light. He even went as far as making calculations, determining that it would take something millions of times heavier than the sun (assuming it was about the same size).

The French scientist Pierre Laplace expanded on the notion, but for years it remained only an interesting—but bizarre—idea. Then in 1916 Einstein published his general theory of relativity. And although he tried, Einstein could not find a solution to the equations of his theory. The first solution came from Karl Schwarzschild, a soldier serving on the Russian front in World War I. When he received a copy of Einstein's paper he was already in a weakened condition from a rare disease he had contracted while serving. In a short time, though, he had solved the equations and sent the solution to Einstein. Einstein was extremely pleased and wrote Schwarzschild. Soon after receiving the reply, though, Schwarzschild died. But his solution has lived on and is still one of the most important solutions.

Although Einstein was pleased with Schwarzschild's solution it disturbed him. It predicted the existence of a bizarre object that had all its mass at a point—a place where space-time was severely distorted. This is the object we now refer to as a black hole.

Einstein tried for years to prove that such objects couldn't exist in our universe. But he did not succeed. Then in 1939 Robert Oppenheimer and Hartland Snyder showed that they could occur when a giant star died and collapsed in on itself. If the star was massive enough (greater than about three solar masses), it could become a black hole.

What would such an object look like if you could observe it up close? First of all, because it absorbs all light that falls on it, it would appear black, and it would be spherical. If you ventured close you would be pulled toward it, but when you got to its surface you would find something strange: you would pass right through it. This is the "event horizon" of the black hole. Once you were through this surface, though, there is no return. You cannot get out—it's a one-way jail. At the center of the black hole is the singularity. This is the remnant of the star that collapsed to produce the black hole; all of the mass of the black hole lies there.

The type of black hole that we will be talking about in relation to the early universe is similar to this, but it did not arise from the collapse of a star. Stellar collapse black holes are all a few miles across. It is impossible to have an extremely small one—say the size of an atom. But such black holes are possible. They may have been created in the big bang explosion. Some of the matter that went out in the explosion no doubt got severely compressed, just as matter does when a giant star collapses. And this matter would have produced black holes—many of them tiny. We sometimes refer to the smaller ones as mini black holes.

In the early 1970s Stephen Hawking of Cambridge University in England discovered something startling about these mini black holes. Examining the region next to their event horizon (mathematically, of course) he found that space was stretched to such a degree that particles and antiparticles literally "popped" out of the vacuum. Some of these particles would be immediately pulled into the black hole, but some would escape, and others would annihilate with their antiparticles, creating radiation, some of which would escape. The net result would be a tiny black hole that emitted particles and radiation. But anything that radiates is hot. This means that mini black holes are hot. Furthermore, if they are hot (giving off energy) they must be

Curvature of space near a black hole.

losing mass and therefore getting smaller and smaller. According to Hawking's calculations they would radiate at an increasing rate as they lost mass, until in the final seconds of their life they would explode.

What would be left? The tiny event horizon would be gone, but not the singularity at the center. But it would now be "naked" (not "clothed" in an event horizon). Physicists had long considered the possibility of naked singularities. Hawkings showed that they could, indeed, exist.

Actually, it turns out that all black holes, regardless of their mass, radiate, and therefore have a temperature slightly above zero. But for large black holes the effect is so slight it would be completely impossible to detect.

One last comment on black holes. So far I have been talking about what is predicted according to Einstein's general theory of relativity. But have we actually seen a black hole? In other

An accretion ring around a black hole (upper left).

words, do we have any good candidates? To answer this we must first look at how we would expect to detect them. We certainly would not be able to detect a typical stellar collapse black hole directly—it's only a few miles across. But we can detect black holes indirectly. Consider a double system, one of the two objects being a black hole and the other a star. Assume the black hole is pulling material from the star. If this were to happen we would have a strong X-ray source next to the larger, presumably visible star. Furthermore, we could easily detect the mass of the X-ray source (to be a black hole it would have to be greater than about three solar masses).

Astronomers have considered this and now have several good candidates, the best of which is one known as CYG–X1. They also believe that the cores of certain types of galaxies (radio galaxies) may be huge black holes.

CHAPTER 6

From Chaos to Creation

We are now ready to look at the creation of the universe. As we discuss the details there may be things that surprise you. One that likely will is the infinitesimally small periods of time we will be dealing with. Also, you will likely wonder how scientists manage to describe things in such detail. It does, indeed, give one a strange feeling, talking about things that happened so long ago (18 billion years). As Steven Weinberg recently wrote, "I cannot deny the feeling of unreality in writing about the first three minutes as if we really knew what we were talking about." The incredibly short times may not seem so strange, though, if you stop for a moment and think about an explosion here on Earth. If you wanted to describe it in detail, right down to what happened centimeter by centimeter as the blast moved outward, you would have to consider some pretty short times.

Before we begin our discussion we should ask ourselves if we are really sure the big bang model is correct. Fortunately, we can feel relatively confident. We have seen considerable evidence in the last few years that it is. Several predictions have been made based on it, and so far most of them have turned out to be correct. That is not to say, though, that it is a completely satisfactory theory. We will see later that it is not. But most scientists are satisfied that its overall features are correct. David Schramm of the University of Chicago recently said, ". . . we can now say with some confidence that the universe we live in is some sort of big bang universe. We no longer discuss what the cosmological model should be . . . . We are concerned now with working out the details of our big bang model."

As we travel back in time to the big bang we notice, particularly as we get close to the event itself, that temperatures and densities are much higher than they are now. Eventually we reach temperatures (or equivalently, energies) much higher than anything we can create on Earth. The early universe was, in fact, the "ultimate" accelerator, far more powerful than anything we have ever built.

In the first moments of creation the universe was a soup of quarks and leptons with a temperature of about 10^{20} K. If we could watch it expand we would see it go through a number of phase changes, just as water does. You are no doubt familiar with the phase changes associated with water: the change from steam to liquid water, and the change from liquid water to ice. This phenomenon is, of course, not particular to water; it occurs for all substances.

One of the first things we notice when watching a phase change in water is that there is a rather sudden change in its physical properties. Liquid water is quite different from ice, and of course steam is different again. This occurs because there is a rearrangement of the atoms.

Another change, perhaps not as obvious, is a change in the symmetry of the water as it passes through a phase. If you could look at the molecules of steam, you would see that they are moving randomly: just as many are moving in one direction as another. There is a symmetry associated with their motion. This is not the case with ice. In one clump of ice the molecules may be all lined up in the same direction, but in a nearby clump they will likely be in another direction. The symmetry was lost upon freezing. We will see later that there is a similar loss as the universe passes through various "freezes," or phase changes. (Incidentally, in talking about the universe we will refer to all phase changes as "freezings.")

Something else worth noting is that when ice at 32°F melts to water at 32°F it releases a large amount of energy. Or, thinking of it in reverse, considerable energy is needed to change water into ice. You are likely familiar with this in relation to your refrigerator. A considerable amount of electrical energy is needed to run it. In all phase changes there is a similar release or absorption of energy. And again we have a similar situation in the early universe.

One of the first questions that comes to mind when talking about the early universe is: How do we look back at it? As we saw in the first chapter, one way is to just look out into the universe—the farther we look, the farther back in time we see because of the finite velocity of light. Unfortunately, we can see only so far—the most exciting moments of the very early universe are cut off from our view because the universe was opaque at this time.

Another way to "look" back would be to build larger particle accelerators, for the early universe was, essentially, a large accelerator. But to approximate the energies of the early universe we would need an accelerator that extended to the nearest stars, so again this route is of little help. A final alternative is to use theory—Einstein's general theory of relativity. It also tells us what the universe was like when it was young. But at the very earliest times the universe was extremely energetic and dense, and eventually general relativity breaks down (no longer gives us correct answers). The reason is that the universe has become so dense and small at this stage that quantum effects are important. To explain it we need a quantum version of general relativity—and we do not have one.

Are we likely to get a quantized version of general relativity in the near future? Michael Turner of the University of Chicago does not think so. But he is convinced that recent theories, called superstring theories that treat particles as tiny strings, will eventually get around this problem. "I think superstring theories are the first attempt that has a real chance," he said. "They are quantum theories that treat gravity on a quantum mechanical basis. All previous theories had the problem that when you tried to quantize gravity you got all kinds of infinities that couldn't be swept under the rug. Whether or not superstring theories are correct is a whole other question. It would be nice if they were correct . . . but I'm not so sure that is so important. Once you get the first self-consistent way of treating gravity on a quantum mechanical basis I think that's such a giant leap that something good will come from it."

Superstring theories do have promise, but it may take several decades to fully develop their potential. In the meantime we have to rely on general relativity, so it is natural to ask: when does it break down? It may sound a little crazy, but the theory is good back to about 10^{-43} second after the big bang. This is such an incredibly short period of time, why would anything before this be important? Oddly enough, it is extremely important—because it's associated with creation itself. And so far we do not know exactly how to deal with it.

Incidentally, the time 10^{-43} second is referred to as the Planck time—named after the creator of quantum theory, Max Planck. Furthermore, the time, or as we usually call it—the era—before 10^{-43} second is called the Planck era.

BACK TO THE BIG BANG

In the next section we will begin a detailed discussion of the events that occurred shortly after the big bang. They will be discussed in the order they occurred. To get a better perspective, though, it is perhaps best to consider how things would look if we were traveling back in time to this event. This would be like running the expansion of the universe in reverse, and indeed this is what will happen if the universe eventually stops expanding and collapses back on itself.

The galaxies are all receding from us now. In the reversed picture they will all be approaching us (and one another). They are so far apart now, though, that it will take an extremely long time before anything important happens. Sixteen billion years will pass before they even get close to one another. If we could

Looking back to the big bang.

watch them in a speeded-up version, where, say, millions of years passed in seconds, we would see a curious phenomenon. Each of the galaxies would look like a Christmas tree with its lights blinking on and off. This occurs because stars are born out of cosmic gas, live for a few million or perhaps a few billion years, then die—in some cases quite dramatically in a supernova explosion.

Eventually, though, all the lights blink off and the galaxies grow into huge gaseous spheres, then finally they merge. At this stage the universe consists of an exceedingly thin gas in an inky black background, but strangely, tiny fluctuations generated by a shock wave that occurred within seconds of the big bang exist within it. As the universe continues to shrink the gas heats and what was once invisible microwave radiation begins to glow. At first it is red, but gradually it turns bright yellow, then a fiery blue-white. Finally, at a temperature of about 3000 K all of space blazes forth like the surface of a brilliant white star. Until now we could see through the thin fog, but all we see now is a blinding white fog. Buried deep in this fog, though, are nuclei—a few are deuterium and lithium nuclei, but most are helium.

The universe is now mostly radiation but as it gets hotter the nuclei break apart, and new particles are created in increasing numbers. First electrons and their antiparticles (positrons) appear, then as the universe gets still hotter, protons, neutrons, and other heavy particles are generated. While this is happening the universe is getting smaller and smaller; finally the protons, neutrons, and other heavy particles begin to touch one another, then as they are forced even closer together something astounding happens: quarks begin spilling out of them. Within minutes the universe is mostly quarks and leptons.

The drama increases as we approach the first second of existence of the universe. Most of the important events occur during this time. We pass through a number of phase changes, similar to the ones we talked about in relation to water, in which the structure of the universe changes abruptly. The forces of nature also change. First the electromagnetic and weak forces merge, then later the other two join them. But all the while the universe is getting more symmetric, and simpler.

As we push back even closer to the big bang we find particles called X bosons, but fewer and fewer types of particles. The universe is getting even simpler: fewer types of particles, fewer forces. Finally we reach the Planck era and all forces become one. And all particles become one. The energy as we pass into the Planck era is an incredible 10^{19} GeV.

Now let's turn things around and look at it as it actually happened. And while we are at it we will fill in the details.

THE PLANCK ERA

The Planck era is the least understood of all the early eras. And with reason—as we saw earlier we have no theory that

Michael Turner.

adequately describes it. Two of the people who are hard at work trying to understand this era are Michael Turner and David Schramm of the University of Chicago. Both were involved in many of the breakthroughs that have been made in recent years in relation to the early universe.

Turner received his Ph.D. from Stanford University in the late 1970s. I asked him how he got interested in the early universe. "When I started out as a graduate student at Stanford," he said, "my original training was in particle physics, but I got discouraged with the job market and took a couple of years leave of absence. I was an auto mechanic and for a while took care of laboratory animals at Stanford Medical Center. Then Bob

Wagoner, who does relativity at Stanford, had an opening and I took it. . . and I learned general relativity and cosmology. Then I went on to the University of Chicago where the 'connection' between particle physics and cosmology was being born and nurtured. This enabled me to use everything I had learned and I found it very exciting. I liked both particle physics and astrophysics, and this enabled me to do both." Turner has made important contributions to our understanding of the origin of matter, neutrino physics and inflation.

David Schramm, who served for many years as the chairman of the Department of Astronomy and Astrophysics at the University of Chicago, got his Ph.D. from Caltech in 1971. From there he went to the University of Texas for two years, then to the University of Chicago. He has won numerous honors, including an invitation (along with several other scientists) to the White House. He has published over a hundred scientific papers, many on the early universe. I asked him what got him interested in cosmology and the early universe. "I have always been interested in questions about the origin of the universe, and cosmological problems," he replied. "I came into astronomy by way of physics, receiving my bachelor's degree and Ph.D. in physics, working on physics problems related to astrophysical situations."

Schramm and Turner are both convinced that an understanding of the Planck era is critical to our understanding of the early universe. But both agree that because of the breakdown of general relativity it is an extremely difficult era to study.

Despite the difficulties we have a rough idea what this era was like. According to John Wheeler of the University of Texas the universe was so small at this stage that quantum fluctuations were cosmic in size. Space and time would therefore be scrambled in a discontinuous chaotic array. He visualizes the universe as a "space-time foam"—similar perhaps to the foam that forms from soap. Stephen Hawking of Cambridge University has shown us that, in addition, every point of space can become an extremely energetic mini black hole. We are therefore led to the picture of space-time as a foam of mini black holes.

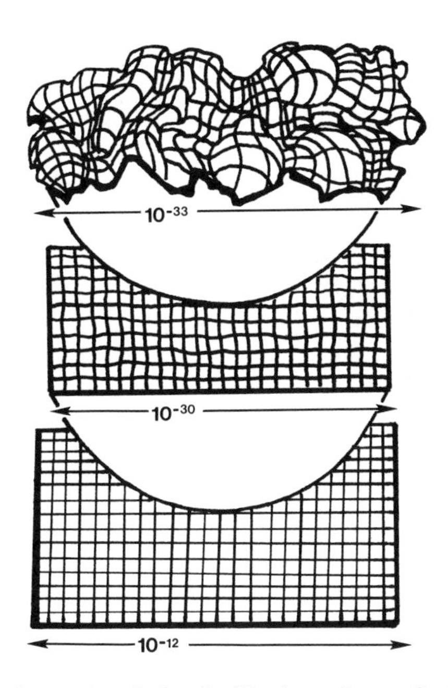

Upper: Distortion of space-time during the Planck era. Lower: Gradual smoothing of space-time. Time is increasing from top to bottom. (Numbers are centimeters.)

David Schramm concurs. "During the Planck era space and time become quantized," he said. "Whether or not this quantization was in our world of 3 plus 1-dimensional space, or in a higher dimensional superstring space, needs to be determined. But with this quantization, all space and time was like a foam of bubbles, popping on and off." The popping on and off he is referring to here is the sudden "popping" into existence of tiny black holes which live only for an infinitesimal fraction of a second before evaporating.

There are, however, a number of prominent physicists who are skeptical of this "foam." Murray Gell-Mann of Caltech is one. I asked Michael Turner about this controversy. "It was never a really well-formulated idea," he said. "The foam is very speculative . . . and has the disadvantage of being vague. Of

David Schramm. (© Patricia Evans.)

course, this may be an advantage—every year you can have a slightly different version of it." He chuckled, then continued. "But now we have the superstring theories . . . perhaps they will have something more meaningful and definite to say about it."

Despite the seeming chaos the universe was actually much simpler at this stage than it would ever be again. There was only one type of particle, interacting through only one type of force—a unification of the four forces we know. To the physicist it was a mathematically beautiful era.

Something you might be wondering about is: How big was the universe at the Planck time? This is an interesting question, but it is not properly phrased. The reason is that we are not sure whether the universe is open or closed. If it is open it will expand forever, and is infinite in extent. Furthermore, it was also infinite at the Planck time. Only if it is closed is this a good question.

We can, however, rephrase it by asking: How big was the "observable" universe at this time? In this case we are referring to the universe we see today, which is a volume of space with a radius of about 15 billion light-years.

We can conclude immediately that the universe was, indeed, much smaller at this time. We merely have to go back in time a few years—perhaps a few billion would be better—and we would see that the galaxies were much closer together. If they were closer together, the space between them would have been less—and the universe would have been smaller.

How small was it at the Planck time? It might be hard to believe but it was not much larger than an atom. Even a trillionth of a second after the big bang it was still only about the size of the Earth. And as late a millionth of a second it was about the size of the solar system.

What about 10^{-43} second before the Planck time? Was the universe a singularity at that time? In other words, was it a region of infinite density—a region where all physics breaks down and we get infinities? According to general relativity we appear to have a singularity at this point. Schramm does not think this was necessarily the case. "To assume there was a singularity 10^{-43} second prior to the Planck time may be totally wrong," he says. He goes on to say that there could be an infinite stretch of time prior to this point. Why? Because what we are actually doing is extrapolating back to time zero. And, of course, we cannot do this because physics has broken down. In fact even the concept of time as we know it has broken down.

THE GUT ERA

At 10^{-43} second the universe entered the grand unified era, an era we also denote by the rather undignified acronym GUT.

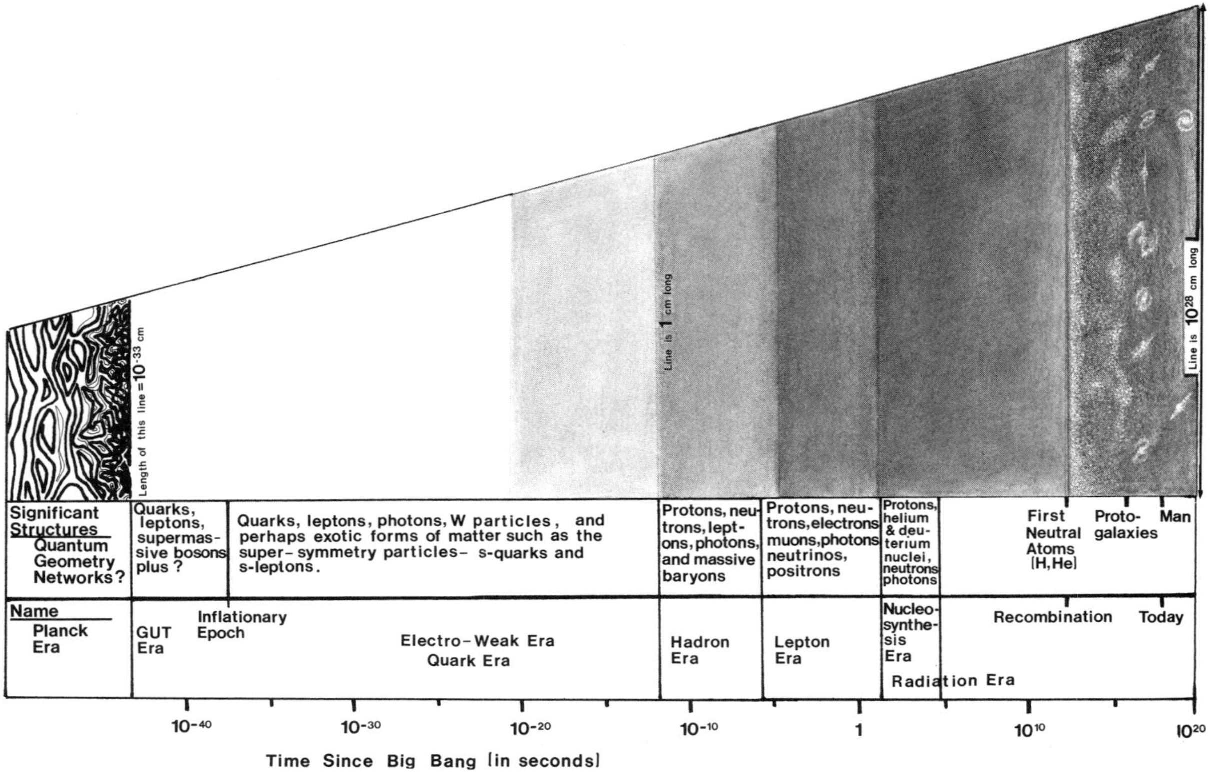

The evolution of the universe.

Why do we call it the grand unified era? Because the grand unified theory explains it; I talked about this theory in the last chapter. According to it, if the energy was between 10^{15} and 10^{13} GeV there will only be two distinct forces of nature: gravity and what we call the strong–electroweak force. This means that as the universe entered the grand unified era, gravity "froze" out, and separated from the other forces.

In the Planck era we only had a single type of particle, but we now have both quarks and leptons. But strangely, we would not be able to distinguish them. As quickly as we identified a given quark it would change to a lepton. According to grand unified theory leptons and quarks are exchanging particles—and in the process switching identity. For lack of a better name the exchange particle is called the X particle (it is a supermassive boson). Other than quarks, leptons, and X particles, there would likely be mini black holes, but they would now be larger.

What about temperature? It would still be exceedingly high—about 10^{28} K. But this is considerably less than the 10^{32} K or more of the Planck era. An apt description of the universe at this point would be a thick soup of quarks, leptons, and X particles, all flying around with exceedingly high velocities. But it would stay this way only until about 10^{-35} second when another freezing would take place.

This freezing is so important it's worth spending a little time describing it (in fact I will later devote an entire chapter to it). To get a feeling for what it was like consider the freezing of water. The first thing we notice as we watch it freeze is that it does not freeze uniformly. Small regicns form first, then gradually grow and merge. We assume that something similar happens at the GUT freezing.

But as the frozen regions merged in the early universe, something important happened. (Or at least we believe it may have happened.) Defects formed along the boundaries. According to calculations there may have been three types of defects, referred to as domain walls, cosmic strings, and monopoles. In each case the defect occurred because there was a sudden change in symmetry around it. You can easily imagine two pieces of ice coming together where the molecules of one piece are all aligned in one direction, and those of the other in another direction. Something similar to this would create a domain wall in the early universe—a two-dimensional defect. On the other hand, if the symmetry suddenly changed everywhere except along a line we would have a cosmic string. And finally, if the symmetry changed everywhere—on all sides of the object—we would have a monopole.

Of these three kinds of defects interest has generally centered on monopoles and cosmic strings. Both would be strange, if they actually existed. Monopoles would be like tiny magnets, but they would only have one polarity—either a north or south pole. And they would be heavy, with a mass of perhaps 10^{16} GeV. Cosmic strings would be equally heavy. A piece a centimeter long would weigh 10^{19} grams.

Interest has also recently centered on another possible event of the GUT era. Alan Guth, of Cornell University, was concerned with some of the problems of the big bang theory when he noticed that most of them could be overcome if the early universe underwent a dramatic and sudden expansion—what we now refer to as inflation. We will look at this idea in more detail in the next chapter.

As the GUT freezing took place the second of the four forces, the strong nuclear force, broke away. The universe then entered the electroweak era.

THE ELECTROWEAK AND HADRON ERAS

The electroweak era is named for the two forces that remain unified. The gravitational and strong forces have decoupled, but the electromagnetic and weak forces remain together. The theory that governs this era is the Weinberg–Salam theory, and the exchange particle is the W particle. In a sense the W particle takes the place of the X particle, as the X particles have now disappeared from the universe because there is not enough energy to create them. And just as the X particle changes quarks into leptons (and vice versa) so too does the W particle change muons into electrons.

The major particles that are now present, then, are quarks, leptons, and W particles, and, of course, photons. But quarks and leptons are now distinct; they do not change into one another. Still, they are energetic. The temperature at the beginning of this era is about 10^{28} K. By the time it is over, though, it will be down to approximately 10^{15} K.

Near the end of the electroweak era an important event occurs: the quark–hadron transition. Above an energy of about 200 MeV the universe is still a soup of quarks, but it is a thick soup; they are as close together as they can get. Then as the universe cools to 200 MeV they begin assembling themselves into "bags." Three go into the baryon bags and a quark and an antiquark go into the meson "bags." And, for the first time, protons and neutrons appear in the universe—roughly in equal numbers.

The energy associated with this transition may seem unbelievably low to you. It did to me the first time I heard about it. After all, we can easily take particles up to 200 MeV; it does not even take an exceedingly large accelerator. Why, then, are we unable to create free quarks? Michael Turner explains: "In the Tevatron at Fermilab the energy per particle is very high, but in order to create this transition you have to reach high temperatures. It's not energy that is so important—it's temperature. In fact the Tevatron is not the place to do this. The place to do it . . . is in a heavy ion collider. You collide uranium with uranium, for example, with enough energy so that when there is a head-on collision, for a moment you create a region where the temperature is raised to 200–300 MeV. Of course, it quickly cools. But you hope that there might be some sign that you have made a transition to quark matter. In fact, there are such experiments going on right now at the BEVILAC at Berkeley. And there is a great deal of discussion about building a heavy ion accelerator at Brookhaven."

Getting back to the early universe we find that it now consists of protons, neutrons, and photons along with a few other particles. Once the hadrons are produced the universe moves into the hadron era. It begins about 10^{-12} second after the big bang. The W particles have now disappeared because there is not enough energy to create them. But the photons are still energetic enough to create proton pairs. There is, in fact, an equilibrium in which proton pairs are being created at exactly the rate at which they are being annihilated.

At this stage there is still roughly the same number of baryons and antibaryons. As the universe cools, though, the equilibrium is broken (this occurs when there is not enough energy to create pairs), and those present began annihilating one another. If the number of baryons in the universe was exactly equal to the number of antibaryons, they would have annihilated one another and nothing would have been left. We know, of course, that this did not happen. There was a small excess of baryons over antibaryons and as a result there was not complete annihilation—a few baryons were left.

THE LEPTON ERA AND BEYOND

At 10^{-4} second the universe entered the lepton era. The temperature was now down to a trillion degrees. Although the photons no longer have enough energy to create proton pairs they are still capable of producing electron pairs. And, as with the protons, we have equilibrium, with photons creating electron pairs at exactly the same rate that electron pairs are annihilating into photons. The major particles that are now present are electrons, positrons, photons, neutrinos, and antineutrinos along with a few protons and neutrons. Soon after this era begins the last of the four forces, the weak force, freezes out and the universe has four distinct forces. When this happens the neutrinos are no longer tied to the hadron–lepton soup. They decouple and fly off. And as the universe expands they continue to cool; their temperature at the present time is 2 K.

We will see later that we have discovered a similar background of photons at 3 K. But we have not yet discovered the neutrino background. Why? The major reason is that neutrinos are so difficult to detect. Most that come to us from space pass right through the Earth without interacting with anything. But it is possible that we will discover them in the next few decades. The difficulties are formidable but scientists are confident; even now they are designing detectors.

As the temperature of the universe drops to 10^{11} K there is still an equilibrium between the electron pairs and the photons. And thinly scattered throughout the radiation are protons and neutrons—about one per billion photons. The neutrons are unstable, however, and soon begin to decay. The number of protons then begins to exceed the number of neutrons. As the temperature approaches 10^{10} K, about one second after the big bang, equilibrium is lost and the electron pairs begin to disappear. Soon the universe is dominated by photons (radiation) and we enter the radiation era.

A particularly important event in the history of the universe is now about to happen: the appearance of the first nuclei. The collisions of protons and neutrons begin creating nuclei of deuterium, but the temperature is still too high for them to remain stable. They are blasted apart almost as soon as they are formed.

Then at three minutes the temperature is down to 10^9 K and the first stable nuclei appear. We talked about the details of this earlier. The sequence is as follows: First a neutron is captured by a proton to form a nucleus of deuterium. Deuterium then absorbs a neutron to make tritium. Tritium decays to helium-3 which is then struck by a neutron to form helium-4. The process occurs very rapidly. In a very short time essentially all the neutrons go into deuterium nuclei and the deuterium nuclei are transformed into helium. In minutes it's all over and the chemical abundance of the universe is firmly set. About one-quarter of its mass is converted to helium; most of the rest remains as hydrogen nuclei (protons). But small amounts of deuterium, helium-3, and lithium are also generated. Nothing beyond lithium was generated, however, because of the gaps at atomic masses 5 and 8.

Nucleosynthesis, as this process is called, is the last interesting event that occurs in the very early universe. The radiation will now just continue to expand and cool. Finally, though, at 10,000 years its temperature will be down to 3000 K. Until now the temperature has been too high for stable atoms to form. Now the nuclei begin to capture electrons and form stable atoms. At this stage the universe is still opaque, as it has been ever since its birth. But as atoms form, the photons decouple from the matter and expand off into space. Suddenly the universe is clear. It's almost as if a shade is lifted and we see out into space for the first time.

The radiation that decoupled from the matter cooled as the universe expanded. And a few years ago it was discovered in an important verification of the big bang theory. It now has a temperature of about 3 K, and as we will see, it is extremely uniformly distributed.

The matter, however, followed a different route. Small fluctuations arose in it and local regions of slightly higher than average density followed. They, in turn, attracted more mass to them and eventually decoupled from the expansion of the universe, and in time, condensed to form galaxies.

REFLECTION

It may seem like I have given a pretty detailed account of the early universe. But the truth is, many of the details have been left out. I said little about inflation, the cosmic background radiation, the helium and deuterium in the universe, and the formation of galaxies. Entire chapters will be devoted to each of these. We begin in the next chapter with inflation.

CHAPTER 7

Inflation

Alan Guth was born in New Brunswick, New Jersey, in 1947. He attended high school in New Jersey, then went on to MIT. All three of his degrees are from MIT; his Ph.D. is in theoretical particle physics. After completing his Ph.D. in 1972 he took a postdoctoral position at Princeton, where he spent three years teaching particle physics and extending his research on quarks. From there he went to Columbia for a couple of years, and then on to Cornell. Seven years after he had completed his Ph.D. he still did not have a permanent position. But with his discovery of inflation all that soon changed.

"SPECTACULAR REALIZATION"

Until late 1978 Guth had paid little attention to cosmology. In fact he admits that everything he knew about it he had read in popular books. But a talk by Robert Dicke of Princeton University started him down a path that eventually led to a significant change in his work and life. Dicke talked about what he referred to as the "flatness" problem of the universe. What do I mean by flatness? To explain it I will have to introduce a number of new concepts. The first, which is called critical density, is not entirely new; we talked about it briefly in Chapter 3. If the average density of the universe is equal to this critical density, the universe is flat; if it is greater than it, space is positively curved; and

cap. 1

Alan Guth.

if it is less, space is negatively curved. The second concept is referred to as omega (Ω). It is the actual density of the universe divided by the critical density. One of the major problems of cosmology is the determination of this number. We are still not certain what it is, but we can say with considerable confidence that it's between 0.1 and 2. (To be perfectly flat it has to be exactly 1.)

The flatness problem, as Dicke pointed out, is related to the fact that omega is so close to 1. It could have any value—as high as 10,000 or as low as 1/10,000—yet it has a value very close to 1, indicating that the universe is almost exactly flat. The significance of this becomes clear when we use the big bang theory to extrapolate omega back to the early universe, say to one second after the beginning. Dicke found that it became 1 to about 15 decimal places. This means that if omega were just slightly greater than 1 at one second after the big bang, it would now be huge—so huge that the universe would have collapsed back on itself long ago. And we would not be here. On the other hand, if omega had been slightly less than 1 at one second after the big bang, it would now be tiny—so tiny that stars and galaxies would not have formed, and we also would not be here. This has convinced some theorists that omega has to be exactly equal to 1.

The problem with this is that the universe appears to have far less than the amount required to make it flat. In fact, it appears to need a hundred times more matter than it has. This deficiency is usually referred to as the "missing mass" (or dark mass).

As a result of Dicke's talk, Guth began to think about cosmology. Still, he found it difficult to take seriously. It was so inexact, so speculative. Then a few months later a colleague at Cornell, Henry Tye, asked Guth if he would be interested in working with him on a problem related to monopoles in the early universe. Monopoles (particles with a single magnetic pole) had been proposed to exist many years earlier (1930s) by Paul Dirac of Cambridge University. In 1974 Gerard 't Hooft of Holland and Polyakov of Russia showed that monopoles also existed in certain types of particle theories. Tye was enthusiastic about the possibility of applying grand unified theory to the early universe. But Guth was unfamiliar with grand unified theories, so Tye explained them to him. Guth quickly realized that the monopoles that 't Hooft and Polyakov had predicted would occur in such theories.

Guth was still reluctant at this stage, however, to get involved in a project in cosmology. But Tye's enthusiasm for the subject helped convince him, and eventually the two men began working together. It did not take long to show that grand unified theory predicted far too many monopoles in the universe—a result that had already been reached by John Preskill of Harvard University. But how would they get around the problem? About this time Steven Weinberg gave a lecture at Cornell on the production of baryons in the early universe. Guth was finally beginning to come around: if someone of Weinberg's stature (he won the Nobel Prize shortly after this) was taking a serious interest in the early universe, there had to be something to it. His enthusiasm for the subject increased. And soon he was hooked.

But in the fall of 1979 Guth left Cornell and moved to the Stanford Linear Accelerator in California. He continued to work with Tye, however, through the mails and by telephone.

His ideas were finally beginning to gel. He now knew a considerable amount about cosmology and grand unified theory, and his strong background in particle physics was becoming invaluable. Together with Tye he came to the conclusion that there was a way of getting around the excess of monopoles predicted by the big bang. If the universe had somehow passed through the temperature where the GUT transition was expected without "freezing," it would enter a kind of "supercooled" state, similar to that which occurs in water. As you likely know it is possible under certain conditions to cool water below 32°F (its freezing point) without ice forming. We refer to this as supercooling. Guth and Tye convinced themselves that something similar had happened in the early universe.

On one of the phone calls Tye suggested that Guth examine the effects of such a supercooling on the expansion rate of the universe. Guth thought about it, but for some reason the request slipped his mind. Then on December 6 he spent the afternoon with Sidney Coleman of Harvard, who was visiting Stanford on a sabbatical. They discussed grand unified theory, X particles, and symmetry breaking. As they talked, things began to click in Guth's mind. He was now sure he was onto something big.

That evening as he left the lab he took his red notebook with him. As he pedaled home he thought about the discussion with Coleman. The ideas were there but they needed to be put in a mathematical form. Excitement about the possibilities continued to build as he thought about them. Finally, about eleven o'clock that evening he pulled out his notebook and wrote "EVOLUTION OF THE UNIVERSE" at the top of the page. He

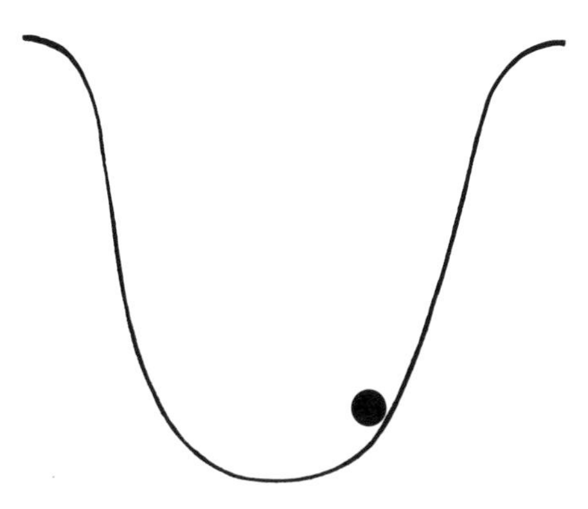

Particle rolling in a trough (an "energy curve").

then began jotting down the relevant mathematical formulas and manipulating them. Thinking about the supercooled state he decided the best way to represent it was as a "false vacuum."

What is a false vacuum? To answer this, let's begin with a particle rolling down a slope (see figure). The particle can roll down and across the valley, but it is trapped by barriers on either side. This means that its energy is restricted by the curve it is following. If we start it at a certain point up the slope, it will roll back and forth; it will, in essence, be in a certain "energy state." The curve that it is rolling on is referred to as its "energy curve."

How do we relate this to the universe? First, we have to ask what the universe consisted of before the GUT transition. There were, of course, energetic particles present so we will describe it as a "high temperature gas of particles." It turns out, though, that grand unified theories also predict that space is at all times filled with what is called a "Higgs field," named for the British physicist Peter Higgs who introduced it.

This Higgs field, it turns out, can also be in various energy states. So, in the same way that our particle was in a particular energy state, so too can we say the Higgs field was in a particular energy state. At the earliest stage, immediately after the

Planck time, it would have been in its equilibrium state, and its "energy curve" would have been similar to that shown on p. 115. But as the universe expanded and cooled it entered a supercooled state and the shape of the energy curve changed. It took on the shape shown in the figure below.

According to Guth the Higgs field got stuck in position A (the false vacuum) at this time, which it is easy to see is not the lowest possible energy. The lowest energy, which corresponds to the true vacuum, resides at B. Eventually, of course, the universe would have to get to this lowest energy state. How would it do this? If we think of the ball at A as a marble in a cup, it seems as if it would somehow have to get enough energy to roll over the hump. When we apply quantum theory to the problem, though, we find something strange: it can "tunnel" through the hump. And when it does, it rolls down to position B—the true vacuum. Physically, this means that the universe began as an expanding bubble of false vacuum (see figure on p. 118); then as tunneling occurred, smaller bubbles of the true vacuum formed within it. These bubbles grew and merged until finally only the true vacuum existed.

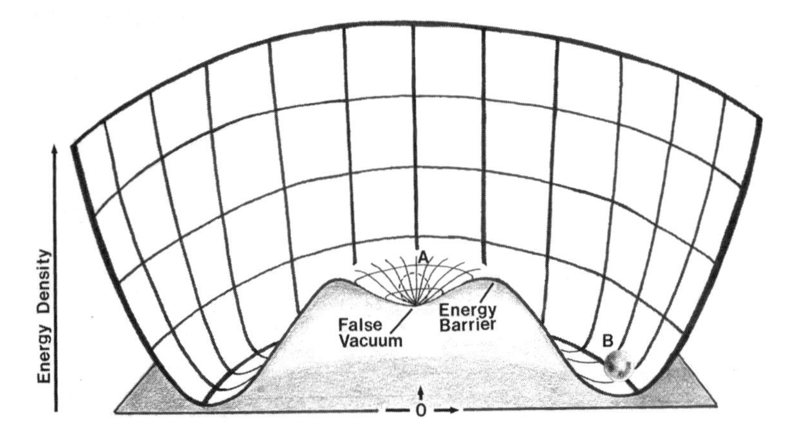

Energy curve showing false vacuum.

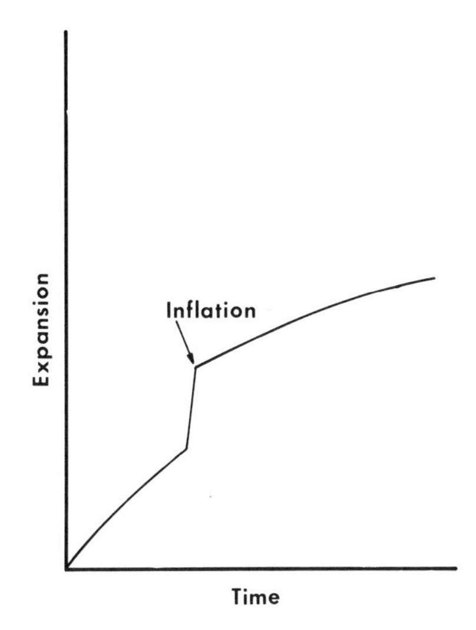

Plot of expansion showing inflation.

But it was what happened when Guth looked at the details of the process that surprised him. In the supercooled state the pressure reversed the effects of gravity. This meant that there was an overall negative pressure that would cause a repulsion. This repulsion would be so powerful it would cause the universe to suddenly balloon out, doubling in size hundreds of times in a tiny fraction of a second (a much faster rate than the big bang). This "inflation," as Guth called it, would begin about 10^{-36} second after the big bang and last until 10^{-34} second. During this time the universe would inflate to about the size of a basketball. And because of this inflation there would be a dramatic release of energy. Just as energy is released when ice melts (this is the energy it took to make it), so too when the GUT "freezing" finally took place at the end of inflation the universe would have been flooded with energy. What happened to this energy? According to Guth it was almost immediately converted into particles and radiation—quarks, leptons, photons, neutrinos—the particles that now make up our universe.

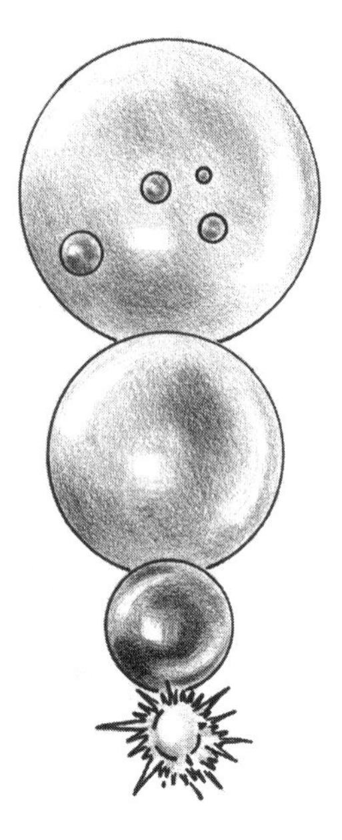

Lower: Expanding universe of false vacuum. Upper: Bubbles of true vacuum forming within false vacuum. Time increases upward.

Guth was amazed at what his equations were telling him. But would this solve the original problem that he had set out to solve: the apparent excess of predicted monopoles in the universe? He soon saw that it did not. But then he remembered Dicke's talk on the flatness problem. A short calculation showed that inflation would, rather miraculously, solve this problem. Inflation would, in effect, flatten the universe. As a simple analogy we could think of a fly crawling on the surface of a balloon. When the balloon is small the fly can easily detect its curvature. But if the balloon suddenly expands hundreds of times, the fly would think it was on a flat surface.

Inflation not only solved the flatness problem, it implied that the universe *had* to be flat. And as we will see later, this, for many, is one of the present difficulties of the theory.

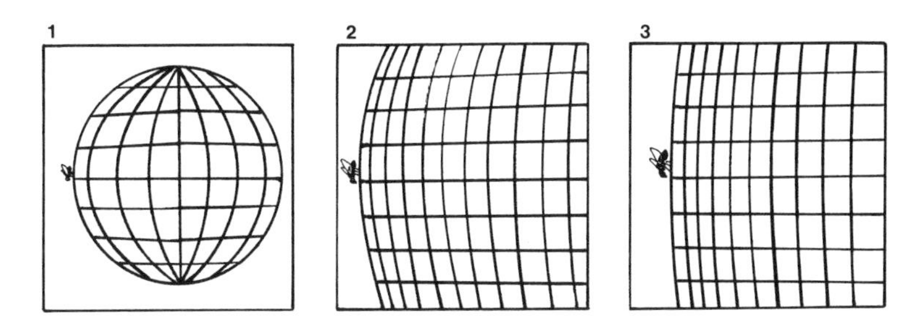

Fly on inflating balloon.

The next morning Guth rushed to the accelerator center. He was anxious to share his discovery with his colleagues. When he got to his office, he pulled out his red notebook and wrote "SPECTACULAR REALIZATION" at the top of the page. He then summarized his results. Later in the day he discussed his calculations with some of his colleagues and was delighted to find that they could find no fault with them. He then gave several talks on the subject.

He was discussing his work with Marvin Weinstein of SLAC one day when the discussion turned to another of the big bang's problems: the horizon problem. It had been shown many years earlier by Wolfgang Rindler that the universe was not causally connected. To see what I mean by this let us assume we spot a galaxy (or better, a quasar) at a distance of, say, 10 billion light-years. Then we turn to the opposite direction and find another at an equal distance. These two objects will then be separated by 20 billion light-years. But our universe is only 18 billion light-years old; this means that a light signal that set out at the time of the big bang from one of the objects still wouldn't have arrived at the other. And because the speed of light is the uppermost velocity in the universe, they could never have been in "communication" with one another. Yet looking at them, they look similar. Furthermore, looking at the radiation around them, we find that it is at a temperature of about 3 K in both cases. How could this be if they were never in contact? The big bang theory cannot explain this. But inflation can. According to Guth, during inflation all regions of the universe were thoroughly mixed, and therefore in communication with one another.

Things were beginning to look exceedingly bright for inflation theory. But the major breakthrough, and perhaps the most important contribution that inflation theory makes, was yet to come. A question that had generally been ignored in relation to the big bang theory was: Where did the energy of the universe come from? Most cosmologists had just tended to shrug and say it was created in the big bang explosion. But to some this was an unsatisfactory explanation.

With inflation theory we did not need to hedge around this question. Inflation produced energy—a tremendous amount. Enough, in fact, to drive the universe to its present state. Even if there was practically no energy initially, after inflation there would be enough to generate the present universe. The universe was, in effect, created out of nothing. Needless to say, this was a dramatic conclusion.

Over the next few months Guth wrote a paper for *Physical Review*. He began with a discussion of the flatness and horizon problems. Then, after briefly reviewing the big bang model, he introduced inflation with the statement, ". . .the universe will continue to cool as it expands, and it will then supercool . . . . Suppose that this supercooling continues down to some temperature many orders below T [the temperature it would freeze at in the big bang model]. When the phase transition finally takes place heat is released." The paper was published in January 1981, five months after Guth had mailed it.

Strangely, the problem that he initially set out to solve—the monopole problem—is not solved by the theory. Furthermore, for all its success, inflation theory—at least Guth's version—had a serious problem. And Guth pointed it out in his paper. There appeared to be no way to smoothly end the inflation. Guth ended it abruptly, which to him (and others) was unsatisfactory.

Before we talk about this problem I should mention that, as

was the case with many other important discoveries, several people were on the verge of this one. Pieces of the puzzle were whirling around in the heads of several physicists about the same time. They just came together more quickly, and more completely in Guth's case. The theory of phase transitions was developed by David Kirzhnits and A. D. Linde of the Soviet Union in the mid-1970s. A theory of bubble formation during phase transitions was also developed independently by Linde at about the same time. The idea that the universe might have expanded at an accelerated rate was suggested in several of their papers. Others who were working on similar ideas at about the same time were Y. B. Zeld'ovich of Russia, Demosthenes Kazanas of the NASA Goddard Space Flight Center, Katsuhika Sato of Japan, and Martin Einhorn of the University of Michigan. But it was Guth with his detailed proof of the cause of inflation and use of it to explain the flatness and horizon problems who really brought the problem to the attention of the scientific world.

NEW INFLATION

Guth began lecturing widely on inflation. He was dissatisfied with its ending, and realized that tied to this was a problem with the way the bubbles in his model merged to give our universe. He worked on the problem with Erick Weinberg of Columbia University, but managed to make little progress.

In 1981 he gave a lecture at Harvard. In the audience was a Junior Fellow of Harvard who had completed his Ph.D. three years earlier under Sidney Coleman. His name was Paul Steinhardt. "I was deeply impressed by Guth's talk," he said. He was aware that cosmology could provide interesting constraints on particle physics, but this was the first time that he realized that particle physics could radically change cosmology. That talk "changed the direction of my particle physics research from formal field theory to 'particle cosmology,'" he said.

Paul Steinhardt.

As excited as he was about the theory, he was nevertheless disappointed. It seemed to have tremendous promise, but how could the inflation be properly ended? Guth had tried, and several others had tried since, but no one had succeeded. Steinhardt and several of his colleagues got together shortly after the talk and tossed around ideas. But no one was quite sure how to proceed.

Shortly after Guth's visit, Steinhardt was talking to Ed Witten, now of Princeton University. "Why don't you look at the electroweak transition?" Witten said to him. "If it also leads to inflation, then the theory is in trouble." He went on to explain that there would be no way you could get the presently known ratio of baryons to photons in the universe if this were the case. The baryon density would be drastically diluted by such an inflation.

Steinhardt went to work and soon showed there were no difficulties. By introducing the proper constraints he was able to

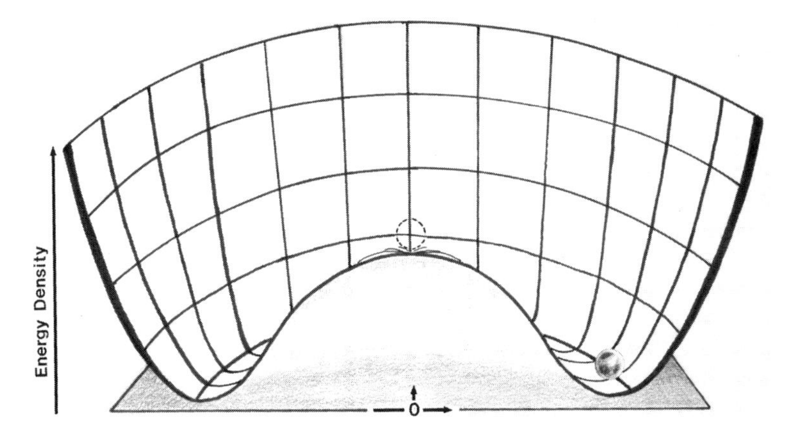

The Steinhardt–Albrecht energy curve.

avoid inflation. "This problem," he said, "was my formal entry into particle cosmology."

Despite his enthusiasm for the beauty of Guth's theory he was still not convinced at this stage that inflation was the way to go. He tried several alternative techniques but finally came back to inflation. If inflation was to work, the major problem, it seemed, was the barrier surrounding the false vacuum, and the way the false vacuum tunneled out through it.

At this point Steinhardt got an assistant professor appointment at the University of Pennsylvania. His first graduate student was Andy Albrecht. Together they tackled the problem and soon found that if instead of a steep barrier surrounding the false vacuum, there was an almost flat region, they could get a smooth ending. The phase transition would take place much slower; they referred to it as a "roll-over" transition.

Also, this "new inflation" model, as it was later called, solved another problem that Guth's model did not—the monopole problem. In Guth's model many small bubbles formed as a result of the tunneling. Our universe was generated by the merging of these bubbles. And because monopoles presumably form along the interface where bubbles merge, there would be a large number of monopoles. In the model of Steinhardt and Albrecht, on the other hand, inflation was associated with a single bubble. This means that our universe arose from a single bubble. And because monopoles only form at interfaces, there would be no, or few monopoles in the observable universe (which lies deep within the bubble). Of course, Steinhardt and Albrecht's model also overcame the flatness and horizon problems, as did Guth's.

While Steinhardt and Albrecht were working on their model, in Moscow, unknown to them, A. D. Linde was working on a similar model. Linde was born in Moscow in 1948 and studied physics at Moscow University. His Ph.D., which was done under the direction of D. A. Kirzhnits, was on phase transitions in the early universe. "At the beginning of 1981," he said, "I felt unhappy because I could see no way to improve the old scenario, and I could not believe that God would miss such a good way to simplify the work of creation." Then late one evening the solution came to him in a flash. "All the details quickly crystallized into a very sharp picture, and I was greatly excited," he said. He got up from his desk and said to his wife, who is also a physicist, "Listen, do not take it too seriously. I must think about it tomorrow, but it seems to me that I have guessed how the universe was created. . . ."

He quickly wrote a paper outlining his ideas, then began to worry. Would anyone take him seriously? "I wrote the text again and again," he said, "trying to answer objections raised by Stephen Hawking and others with whom I had discussed my ideas." Finally he submitted the article to *Physics Letters*, and at about the same time circulated some preprints.

Steinhardt and Albrecht had been working on the problem for about eight months when they received one of Linde's preprints. They were shocked. "We thought we were alone on our approach, and we were worried about many as yet unresolved issues," said Steinhardt. Looking at the title of Linde's paper, and quickly scanning through it convinced them that they had

Andrei Linde.

been "scooped." But when they read the paper in detail they found that, although Linde had arrived at the same conclusion they had, in other words that a flat-topped barrier overcame the problem, Linde had approached the problem from a completely different point of view. Furthermore, he had ignored the "unresolved issues" that they were currently worried about. Steinhardt and Albrecht therefore decided that it would be fair to write up their results, and they did so within a few days of receiving Linde's paper. They felt that it was clear that their work had been done simultaneously and was independent of Linde's.

When Linde received the preprint by Steinhardt and Albrecht he was surprised. His initial reaction, he said, was that he was glad he was not alone in the approach, and that now it would be easier for people to believe that what he had done was not crazy. "But I was surprised that they independently suggested essentially the same scenario that I had, after I had distributed my preprints, but I realized that it was possible. I was not very happy that I had lost several months writing and rewriting my paper."

Both groups are now credited with independently making the discovery.

But even new inflation had problems. The major one was that it did not explain how galaxies formed. It had, for many years, been assumed that small fluctuations somehow formed in the early universe, and eventually gave rise to the galaxies. But exactly what caused these fluctuations was unknown. Did inflation play an important role? Steinhardt believed that it did. Working with Michael Turner and J. Bardeen he tried to calculate what type of fluctuations would be generated by inflation. But we will leave this part of the story to later.

VARIATIONS ON NEW INFLATION

Because problems remained, several variations on new inflation were soon put forward. Some of them were based on alternate grand unified theories, but most were based on new physical ideas that were emerging. A variation that is still getting a lot of attention is based on supersymmetry. Supersymmetry is a theory devised by Julius Wess of Karlsruhe University and Bruno Zumino, who is now at the University of California, in the mid-1970s. They looked at the two basic types of particles in the universe, fermions and bosons, wondering if they could devise a way of changing (mathematically) one type into the other. If so, the universe would be much simpler. And lo and behold they found that it was possible. The major triumph of the theory, though, came a couple of years later when Zumino and Stanley Deser of Brandeis University, and independently, Sergio Ferrara of Italy, Daniel Freedman and Peter Nieuwenhuizen of the State University of New York at Stony Brook showed that general relativity (gravity) could be included in the theory. The newer version is called supergravity. This was the first time scientists had been able to unify gravity with the other

forces of nature and it was considered to be a major breakthrough. In time, though, supergravity was found to have problems. (Because they are not important in our story we will not go into them.)

Coupling supergravity and supersymmetry with inflation leads to several different theories. And although theories of this type have been helpful, they still do not seem to be the final answer.

"Should we be optimistic about supersymmetric inflation?" wrote Steinhardt in 1986. Answering his own question, he said, "It's hard to say. The route has proven to be much more difficult than was originally proposed. On the other hand, each time an obstacle has been overcome, some extra benefit has come for free."

More recently, a number of the problems of supersymmetry and supergravity have been overcome by an even broader theory called superstring theory. In this theory particles are represented by tiny vibrating strings. And, as you might expect, it was not long before the first superstring inflation theory was formulated. It also appears to have considerable promise.

Michael Turner has recently teamed up with Joseph Silk of the University of California to examine the possibility that there was more than one inflation associated with the GUT transition—perhaps two, or even more. Their main concern was the formation of galaxies. As I mentioned earlier, this is still one of the major problems of inflation theory.

CURRENT STATUS OF INFLATION THEORY

Inflation theory, most cosmologists agree, is an excellent idea that helps us avoid many of the problems of the big bang model, but it is still not out of the woods. I talked to several well-known cosmologists about it and got various opinions. According to Michael Turner, "There are no fundamental problems with implementing inflation. Probably the biggest problem

the present is that there is no particle physics model that in a convincing way implements it." He went on to say that he believes it has about a 20% chance of being correct. When I expressed some surprise at what seemed to me to be a low number he said, "If you take your ordinary garden-variety beautiful theory . . . having a 20% chance of it being correct is actually phenomenal." Turner said he was particularly pleased with the theory because it makes predictions that are becoming testable. "One example," he said, "is the prediction that the universe should be flat." Nodding his head he added, "Yes . . . I'm quite high on inflation."

David Schramm agrees with Turner in general, but has some reservations with present theories. "I believe that some sort of inflation probably occurred in the early universe," he said. "However, the details of how this inflation occurred remain to be determined. . . . It is clear that almost any grand unified theory can yield inflation. What is not clear is how to get this inflation to be consistent with such things as the fluctuation spectra needed to make galaxies. I believe there are details that remain to be worked out."

Edward Kolb, head of the astrophysics group at Fermilab, who has also worked on inflation, says, "Some sort of inflation is correct, but I don't believe it has anything to do with phase transitions." Jim Peebles of Princeton University says: "Inflation is a beautiful idea . . . unfortunately it is lacking in predictions."

In an effort to get a view from an observational astronomer (as opposed to a theorist) I talked to John Huchra of Harvard–Smithsonian. He expressed concern. The major problem inflation has, he said, is that it predicts a flat universe. He finds it extremely difficult to believe that enough "dark matter" will be found to make the universe flat. "My observations don't appear to support inflation," he said.

Gary Steigman of Ohio State University says he tends to agree with Huchra. He is also concerned with the fact that inflation predicts a flat universe, but is quick to point out that, regardless of inflation theory, the flatness problem is a serious

problem. "Inflation is one answer," he said, "but I'm not convinced it's the only answer. There is a problem independent of whether you believe in inflation or not. I've argued in recent years that there is no good evidence that omega is equal to 1, and so I think I side with Huchra on the side of skepticism."

Steigman went on to point out that there have been studies based on galaxy counts that indicate omega is equal to 1. "But the large body of data does not support this conclusion," he said.

Despite the fact that Linde formulated the new inflation model, he is not entirely satisfied with it. Recently he has become interested in a variation that is quite different. "In my opinion," he says, "there exists a much simpler scenario which is not based on the theory of phase transitions. . . ." He refers to his new theory as the eternal chaotic inflation theory. "According to this theory the universe has no end and may have no beginning. The universe in this case consists of a large number of self-reproducing mini-universes inside of which all possible types of vacuum states are realized. Despite its weird qualities it is basically simpler than new inflation."

It seems that in general, most cosmologists agree that inflation is a good idea and the way it overcomes the problems of the big bang model is excellent. But most agree that problems remain.

CHAPTER 8

Mystery of the Cosmic Mirror

One of the major mysteries of the universe is why there is so little matter in it. It is, in fact, so empty that Willem de Sitter, when chided about proposing an "empty" cosmological model, quipped, "Well, the universe is practically empty." Despite this apparent drawback, de Sitter's model was taken seriously for many years because, even then, it was known that the universe had very little matter in it. Stars are separated on the average by about 5 light-years (a light-year is the distance light travels in a year). They are so far apart, in fact, that when two galaxies collide they pass through one another without a single collision between stars. And the space between the galaxies is even emptier. Galaxies in clusters usually have a small amount of intergalactic matter dispersed throughout them, but between the clusters there is literally nothing.

Strangely, though, it's not the fact that the universe is so empty that has puzzled scientists in recent years. It's the fact that it has any matter in it at all. According to recent discoveries it should be empty.

To understand why this is so we have to go back to the late 1920s. Paul Dirac of Cambridge University in England was trying to solve one of the major problems of quantum theory when he discovered that his equations predicted a particle similar to the electron, but of opposite charge. At first he thought it had to be the proton. But the proton was much heavier than the electron. The only alternative, it seemed, was that it corresponded to a particle that had not yet been discovered.

And, indeed, his prediction was soon borne out. In 1933 Carl Anderson of the California Institute of Technology found the new particle, calling it the positron. But this new particle was no ordinary particle. When brought together with the electron, both it and the electron would disappear in a burst of energy. They would, in essence, annihilate one another.

Did this also apply to other particles? Did the proton, for example, have a "cousin" that would produce annihilation? It did. But it was many years before it was discovered. Not until 1955 did Owen Chamberlain and Emilio Segrè of the University of California first detect it.

Scientists were soon convinced that all particles had antiparticles. And many more were soon discovered. We now believe that to each particle there corresponds an antiparticle, and when any particle collides with its antiparticle they annihilate one another. The antiparticles of literally all known particles have, in fact, now been discovered. They can be created in laboratories along with their matter particles in high-energy collisions. Indeed, most were discovered in such collisions when both particles emerged.

Once scientists realized that antimatter could exist they began to wonder how much of it there was in the universe. Was it possible that antistars or even antigalaxies could exist? And if so, could we detect them? The answer to the first question is yes; in theory they can exist. But we would have difficulty distinguishing them from ordinary stars and ordinary galaxies. This is because both emit photons (the photon is its own antiparticle) and therefore they would look the same.

HOW MUCH ANTIMATTER IS THERE IN THE UNIVERSE?

Is it possible that a large part of the universe is made of antimatter? In 1976 Gary Steigman, who was then at Yale University, wrote a review article summarizing the evidence. He looked carefully at the entire universe, beginning with the solar system and continuing through the stars and galaxies. He pointed out that there was no evidence for antimatter in the solar system. We have, after all, he said, sent probes to the moon and several of the planets. If any of the objects they landed on were made of antimatter the probes would have quickly disappeared in a burst of energy. And, of course, we know of no such bursts. Furthermore, the solar wind "blows" throughout the solar system, and since it consists of matter particles, if it encountered any antimatter there would be a tremendous explosion. And again we have seen no such explosions.

What about beyond the solar system—among the stars? Again there appears to be little hope for antimatter even there. The best evidence that this is the case comes from cosmic rays. We are still uncertain where cosmic rays come from, but it is reasonable to assume that most come from somewhere within our galaxy. Cosmic rays are high-energy particles (primaries) that strike the particles of our atmosphere, generating showers of "secondary" particles. Near the surface of the Earth we detect mainly these secondaries. To detect the primaries we have to get above our atmosphere (or at least most of it).

A small number of antiparticles are, indeed, found within cosmic rays. But they are likely generated in the collisions of the primaries with the molecules of our atmosphere. Indeed, the number observed is not inconsistent with the number predicted in such collisions. Furthermore, most of the antimatter we see is in the form of positrons. They are the antiparticles we would expect to be most common in such collisions. If these particles were coming from space we would expect them to be accompanied by antiprotons, and other antimatter in the form of heavier antiparticles. In fact, we would likely see a few antinuclei. But we do not. Why? The only reasonable answer is that there is not much antimatter out there. If we did detect some antinuclei, say some antihelium, we could feel confident that there was antimatter in space. But so far we have not.

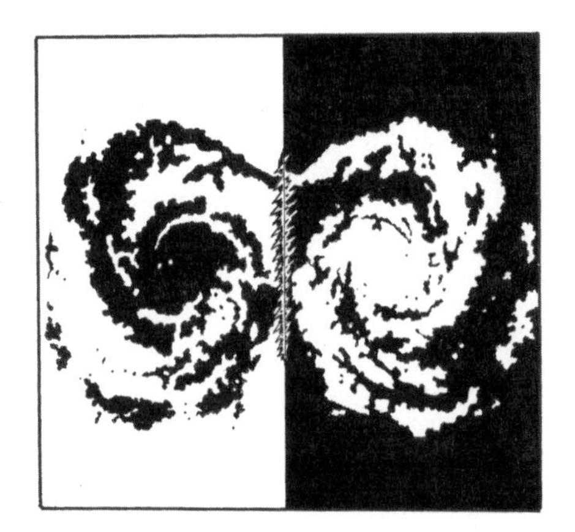

Collision of a galaxy composed of matter with one composed of antimatter.

Another way we might detect antimatter is through the gamma rays that would be emitted in an antimatter–matter annihilation. Consider, for example, the collision of a star and an antistar (a star made of antimatter), or perhaps better, a galaxy and an antigalaxy. Certainly, if two such objects came together and totally annihilated one another there would be an awesome explosion with the release of a tremendous amount of energy in the form of gamma rays. In theory, the collision of two galaxies would produce the most powerful explosion possible in the universe. What could be greater than all the matter and antimatter of two such objects being converted into energy?—only the big bang itself.

Calculations have shown, though, that it is impossible for a galaxy and an antigalaxy to collide with enough force to totally annihilate one another. In practice, as they approached one another the outer regions of the two objects would interact creating a powerful explosion that would quickly blow them apart. But even if they did not totally annihilate one another, if two such objects did collide they would be an exceptionally strong source of gamma rays.

Have we seen such collisions? We do, in fact, observe galaxies colliding with one another, and they are strong sources of radiation—mostly radio waves. But the radiation we receive is consistent, not with matter–antimatter annihilation, but with the collision of two hydrogen clouds. And literally all spiral galaxies have hydrogen strewn throughout them. When the two galaxies collide the stars do not interact (their separation is too great) but the hydrogen clouds do, and in the process they give off radio waves.

In 1972 a diffuse background of gamma rays was discovered, and there was speculation that it might come from matter–antimatter collisions. But so far we have not found any evidence that this is the case. It seems more likely that it is just a residue from known gamma ray sources such as the Crab Nebula.

I mentioned earlier that we cannot distinguish matter and antimatter stars (or galaxies) by examining the photons they emit. There is a way, though, that we could distinguish such objects by observing particles that come from them. And that is by studying the neutrinos (or antineutrinos) they emit. A star composed of matter, for example, emits mostly neutrinos, whereas one composed of antimatter would emit mostly antineutrinos. If we had a "neutrino telescope" we might be able to distinguish such objects.

It will likely be a while, though, before we have such an instrument. The neutrino is an extraordinarily elusive particle. It has no charge, no spin, and according to most indications it has no mass (there is the possibility it has a small mass—we will talk about this later). Furthermore, most neutrinos that strike the Earth pass right through it without interacting with anything. We are, however, able to detect them in the laboratory. So, although it may be a ways off, scientists will likely develop a "neutrino telescope."

So far we have been talking about detecting antimatter by observing it directly or observing its effects, but there is also an indirect way of determining its presence. If we could devise a mathematical model of the universe—a cosmology that predicted the existence of antimatter, and was consistent with all observations—we would have a strong indication of its presence. And indeed several antimatter cosmologies have been devised. Most assume that the universe began with equal amounts of matter and antimatter. They then explain, or perhaps I should say, attempt to explain how it remained separated so that an annihilation catastrophe was averted. R. Omnes of the University of Paris gets around this problem by assuming that the initial temperature was so high that particles and antiparticles underwent a phase transition that kept them apart. He has modified his theory considerably over the years but has never come up with what might be considered an acceptable mechanism for avoiding the annihilation catastrophe.

Hannes Alfvén and Oskar Klein have presented a model in which creation begins with an immense sphere containing a uniform distribution of matter and antimatter. They call this sphere the *metagalaxy* (it is about a trillion light-years across). According to their theory gravity gradually pulls particles and antiparticles toward the center of the metagalaxy, and as they move inward they fall faster and faster, gaining momentum. Soon antiparticles and particles begin to collide and annihilate, and radiation begins to accumulate. Finally a tremendous outward pressure is built up and the inward fall is stopped and reversed. This reversal is responsible, according to them, for our present expansion. It's an interesting and imaginative theory but it also has problems—serious ones. First of all, it is in conflict with general relativity (particles do not attract antiparticles in general relativity); furthermore, it fails to properly account for the cosmic background radiation.

More recently Floyd Stecker of NASA has devised a model in which separate domains of matter and antimatter are created. His universe resembles a honeycomb. But until proof comes that the universe is indeed constructed in this way few are likely to take him seriously.

So again we have a dead end. We cannot construct consistent cosmologies that predict antimatter in the universe. Does this mean there is no antimatter? We still cannot go as far as saying that. The evidence against it is certainly strong, but it is possible that there are regions of antimatter well separated from matter. They would have to be separated on a scale larger than galaxies. Most galaxies are in clusters, and there is intergalactic matter between the individual galaxies. If they were not all composed of matter (or antimatter) the cluster would be a strong gamma ray source. This tells us that they would have to be separated at least on the scale of clusters of galaxies. It is possible, though, that there are clusters of galaxies in a remote part of the universe that are composed of antimatter. In fact, it is even possible—but quite unlikely—that half of the universe is antimatter.

ANTIMATTER IN THE EARLY UNIVERSE

Let us turn to another question. If the present universe has no antimatter, is it possible that early on it contained some? This, it turns out, is quite a different question. Most scientists working in the field are convinced that when the universe was created it had equal amounts of matter and antimatter. There are several reasons for their conviction. The most obvious one relates to the extreme temperatures at that time. At such high temperatures all particles were undergoing extensive collisions, and we know such collisions create particle–antiparticle pairs. With so many collisions occurring there must have been a large amount of antimatter. In fact, because an antiparticle is created each time a particle is, there should have been equal amounts of the two. Of course, the universe eventually cooled to temperatures below those needed to produce such pairs. Nevertheless, initial temperatures were easily high enough to produce them, and therefore the early universe should have been "symmetric."

Another argument for equal amounts of matter and antimatter comes from the laws of physics. For the most part they show no preference for matter over antimatter. If you exchange particles and antiparticles in reactions the same laws still apply. (We will see later, though, that this is not strictly true.)

But if the early universe, say before about 10^{-35} second, did contain equal numbers of particles and antiparticles we have a serious problem. As cooling occurred each particle would eventually find an antiparticle and annihilate. In fact, all particles and antiparticles would annihilate and the universe would end up empty. Yet the universe has considerable matter in it. Why is it not empty? And why is there only matter left over? Why, for example, is there not just antimatter, or some combination of matter and antimatter?

It is possible, of course, that our ideas are all haywire and the universe was created with just matter in it. But Leonard Susskind of Stanford University and D. Dimopoulos of the University of Chicago challenge this. In their article "Baryon Number of the Universe" they quote Einstein: "If that's the way God made the world then I don't want to have anything to do with Him."

Of course, if the universe did initially have equal amounts of matter and antimatter, in other words if it was symmetric, then there had to be some sort of mechanism for leaving it as we see it today. And indeed cosmologists now believe they know how this happened. But before I can talk about it I will have to introduce some new concepts.

The first is parity, which refers to the mirror image of a process. The conservation of parity says that the mirror image of any particle interaction is also a possible one. In the figure, for example, we see the decay of a muon to an electron, a neutrino, and an antineutrino. The electron is assumed to be spinning in a left-hand direction. In the mirror image the process is the same except that the electron is spinning in a right-hand sense. Is the mirror image process possible? Indeed, it is, but it does not occur nearly as often as the original one. Left-hand electrons appear more than 1000 times as often as right-hand ones. Parity is therefore not conserved.

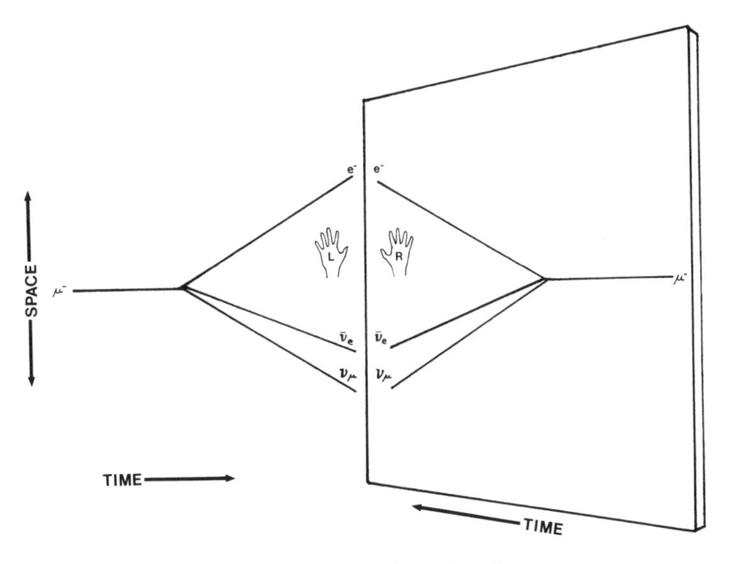

The decay of a muon and its mirror image.

The discovery that it is not conserved came in 1956. Let us take a moment to consider it. The codiscoverers, Chen Ning Yang, who is now at Stony Brook, and Tsung-Dao Lee of Columbia, first met when both studied at the National Southwest Associated University in China. Yang came to the United States in 1945 on a scholarship to work with Enrico Fermi, who was then at the University of Chicago. Fermi had a few years earlier produced the first sustained fission reaction. Yang obtained his Ph.D. under him in 1948. Lee also attended the University of Chicago, receiving his Ph.D. in 1950. He worked under Edward Teller, the "father of the hydrogen bomb."

Yang and Lee met again at the Institute for Advanced Study at Princeton in 1951, where they began working together. Lee left for Columbia in 1953 but they continued their collaboration. They began their study of parity by examining the decay of K mesons. There appeared to be two different types of K mesons that decayed in two different ways. But aside from this they were identical. Was it possible that they were actually the same particle? This could only be the case if parity was not conserved. But if this was true it meant that nature could distinguish a right-handed spinning particle and a left-handed spinning one. Surely this was not possible. Or was it?

After studying the interaction for a while Yang and Lee convinced themselves that parity conservation was indeed violated, at least in the weak interactions. Their assertion was checked within months by Madame C. S. Wu of Columbia University, and it proved to be correct. The scientific world was stunned: nature could distinguish right-handedness from left-handedness. Yang and Lee shared the Nobel Prize in 1957—only a year after their prediction.

So parity (P) was not conserved in the weak interactions. But scientists soon discovered that if they coupled parity with charge conjugation (C), a process in which the particles are replaced by their antiparticles (and vice versa), conservation was satisfied. This new process, referred to as CP conservation, was considered to be universally valid. Then in 1964 James Cronin of the University of Chicago and Val Fitch of Princeton University began looking at neutral K mesons (K^0). According to CP conservation the decay of K^0 into two pions would be disallowed. But Cronin and Fitch observed that about one in 500 did decay into this state. This meant that CP conservation was also violated. And in 1980 Fitch and Cronin were awarded the Nobel Prize for the discovery.

Was there any way to salvage a conservation principle? There was also the possibility of time reversal (T), in other words, reversing the direction of time in the process. Coupling this with CP gives CPT, and it is now believed that CPT is indeed conserved.

But what has this to do with how the early symmetric universe evolved into one containing only matter? It turned out that CP violation was the key. Within a year after the discovery

Andrei Sakharov of Russia showed that CP violation could be used to explain our present universe.

SAKHAROV

Andrei Sakharov was born in 1921 in Moscow. His father was a well-known physicist and the author of many physics texts and popular science books. Sakharov spent his childhood in a large communal apartment shared with many relatives. He did not attend public elementary school, but was taught by his parents at home. Because of this, when he later attended public schools he had difficulty relating to his own age group.

He graduated from high school in 1938 and enrolled in the physics department at Moscow University. Despite disruptions in his studies caused by the war he graduated in 1942, then took a job as an engineer in a large factory. It was during this time that he wrote several articles on theoretical physics and sent them to the University of Moscow. Officials there were so impressed they recommended that he apply to a graduate school. And in 1945 he entered Lebedev Institute of Physics. Nobel Prize winner Igor Tamm was assigned to him as advisor. Sakharov was pleased, later saying that Tamm had a tremendous influence on his career.

In 1948 he began working with Tamm's group developing the hydrogen bomb. "We were all convinced of the vital importance of our work for establishing a worldwide military equilibrium," he wrote. But gradually his views began to change. He began to think about the horrors that use of the hydrogen bomb and other nuclear devices would bring. His guilt about his role in developing it mounted until finally he felt he should do something to compensate.

In the 1950s he began campaigning for a halt or limit to the testing of nuclear bombs. This was a brave action on his part. Almost anyone would have been quickly thrown in jail. But because he was already well known and had made important contributions to the country he was left alone—at first. But he continued to be a thorn in the side of the government as he began campaigning for human rights.

In 1968 his essay "Progress, Co-existence and Intellectual Freedom" was published abroad. "This was the turning point in my life," he said. The essay got worldwide publicity and many Soviets began looking to him for help and leadership in their fight. By the 1970s he was heavily involved in human rights and the fight against political repression. He was soon barred from all classified material, and pressure on both himself and his family increased.

The final blow came in 1975 when he was awarded the Nobel Peace Prize. His government refused to let him go to Oslo to accept the prize (it was accepted by his wife in his place). His acceptance speech, "Peace, Progress and Human Rights" firmly echoed his beliefs. But it was too much for the Soviet government; they soon stripped him of all his official Soviet awards and banished him to Gorky, a city in northern USSR. He tried to continue his scientific work there but was not allowed to communicate with anyone, and was literally cut off from the scientific community. Only recently was he allowed to return to Moscow.

Sakharov has had so much publicity in regard to human rights that his scientific work has been overshadowed. Despite his extracurricular activities he did make many important contributions to physics—plasma physics, cosmology, field theory, and elementary particles in particular. One of his most important cosmological contributions dealt with the matter–antimatter asymmetry of the universe.

In 1966 Sakharov asked himself the question: If the laws of physics are symmetric with respect to matter and antimatter and it is likely that the early universe consisted of equal amounts of matter and antimatter, how did the universe end up with only matter in it? The explanation that God made the universe this way "right from the beginning" did not appeal to him. He began looking for a scientific answer, and soon realized that CP non-conservation was the key. Particles and antiparticles could be produced at slightly different rates if such were the case. But he soon realized that violation of CP (and C) conservation was not enough. Something else was needed.

To understand what else he required I need to introduce what is called "baryon conservation." We saw in an earlier chapter that the heavy particles of the universe, particles such as protons and neutrons, are known as baryons. Physicists have found it convenient to label baryons and other particles with various quantum numbers; in practice these numbers are little more than a bookkeeping device. One of them is called baryon number (B). Protons, neutrons, and all other baryons are given the baryon number B = 1. Antiprotons, antineutrons and other antibaryons are given the baryon number B = −1. All other particles are assigned B = 0.

Now for baryon conservation. It implies that in any interaction B remains constant. Whatever B is before the interaction, it must be the same after. And for years scientists were convinced that B was conserved. There was, in fact, a strong reason for their belief. If it was not conserved the proton would be unstable—it would decay. And everyone was confident that this did not happen. If it did decay, and had a lifetime of less than 10^{16} years physicists would be able to detect radiation coming from our bodies. And if this were the case, cancer would be rampant.

Nevertheless, Sakharov showed that the nonconservation of baryon number would be needed to leave the universe with its preponderance of matter. Furthermore, he specified that the universe must go from a state of equilibrium to one of non-equilibrium. With these two conditions and CP violation, he said the universe could end up the way we presently see it.

But if baryon number was not conserved the proton had to decay, and indeed he calculated its expected lifetime, getting a large but finite number. The surprising thing about Sakharov's proposal, though, is that it was so far ahead of its time. And because of this it was generally ignored outside of the Soviet

Union. For over ten years the proposal lay dormant. Finally, though, with the advent of grand unified theory it was rediscovered.

MORE RECENT THEORIES

The scientific world was not ready for Sakharov's ideas in 1966. But between 1966 and 1978 several important advances in particle physics brought the problem once again to the attention of physicists. The first of these advances was a successful unification of the theory of the electromagnetic field (quantum electrodynamics) and the theory of the weak interactions. Steven Weinberg and, independently, Abdus Salam showed that they could be brought together. Although it was not a completely satisfactory unification in that it had many unspecified parameters, it showed that the electromagnetic and weak fields could be mixed successfully.

A few years later an excellent theory of the strong interactions emerged. It is now referred to as quantum chromodynamics. In this theory the strong interactions were assumed to be due to particles called gluons that moved back and forth between quarks.

The next logical step was the unification of the electroweak theory with quantum chromodynamics. And several such attempts were soon made. The first, which came in the mid-1970s, was due to Howard Georgi and Sheldon Glashow. One of the predictions of their theory was a particle referred to as X that could change leptons to quarks (and vice versa). But if X particles existed the proton had to decay. It was, in fact, predicted to have a lifetime of about 10^{31} years.

But if it takes 10^{31} years for the proton to decay how could we ever measure it? The universe is only about 10^{10} years old. Fortunately, we can get around this. If we assemble, say, 10^{34} protons, one of them should decay every few days. And, as it turns out, 10^{34} protons is not an overwhelming number; they could easily he housed in a small building. Another advantage of such an experiment is that protons of one material are the same as protons of another. We can therefore use relatively cheap materials such as water or iron. Using such materials several experiments have been set up: one in an old gold mine in India, one in a tunnel under Mont Blanc on the border of Italy and France, one in an old salt mine in Ohio, and at several other locations. So far, unfortunately, no one has caught a proton in the act of decaying (there was one report that now appears to be false). But many scientists are convinced they will eventually detect the decay.

Anyway, with the development of the above theories it was inevitable that someone would begin applying the ideas to the early universe. And, indeed, in 1978 Motohiko Yoshimura of Tokohu University in Japan proposed that CP and baryon violation could yield our present universe. In his paper he states, "The essential point of my observation is that in the very early, hot universe the reaction rate of baryon number nonconserving processes, if they exist, may be enhanced by extremely high temperatures and high density." He goes on to show that such high temperatures would produce more quarks than antiquarks. He then calculated the magnitude of the difference and got a number that did not seem unreasonable.

It was soon discovered, though, that there were problems with Yoshimura's approach. He overlooked an important part of the problem, and if it had been included his universe would have ended with no matter. Nevertheless, his paper was important in that it introduced a number of significant ideas.

Interestingly, at almost the same time that Yoshimura was working on the problem, others were also beginning to work on it. Two groups in the United States had gotten into the act; one consisted of Leonard Susskind of the Stanford Linear Accelerator at Stanford University and Savas Dimopoulos of the University of Chicago, and the other of D. Toussaint, S. B. Treiman, and Frank Wilczek of Princeton University, and A. Zee of the University of Pennsylvania.

Susskind did his undergraduate work at CUNY, then worked for several years as a plumber in the Bronx. In the early 1960s he decided to return to college, and in 1965 received his Ph.D. from Cornell. I asked him about his and Dimopoulos's work. He said that Yoshimura's paper was brought to his attention by Bob Wagoner of Stanford. "When Bob asked me about it I was really ignorant of cosmology," he said. "But I told him I would look at it. I had known about the problem for several years, but had never given serious thought to it." He soon realized that Yoshimura had made a mistake: he had not assumed that the universe would go from equilibrium to nonequilibrium. He and Dimopoulos then showed that in an expanding universe baryon violating processes would likely be out of equilibrium if they occurred sufficiently early while the expansion was rapid. We saw earlier that nonequilibrium was one of Sakharov's conditions. All of the key ideas were now there but Dimopoulos and Susskind did not give a detailed process for the development of the baryon excess. This was supplied by Toussaint, Treiman, Wilczek, and Zee.

Frank Wilczek described to me how he and his group became interested in the problem. "The immediate stimulus," he said, "came from some rather distant ideas. At the time I was fascinated with the idea of black hole radiation—essentially just as an interested layman. Also, there was much excitement around Princeton about instantaneous vacuum fluctuations called instantons." Wilczek said he was impressed with the work, but appalled by the approximations that were being made. "I couldn't get into the spirit of the work," he said. But earlier he had worked on a similar problem related to gravity and soon he became interested in it again. "It was natural to think about black holes and the early universe in this connection," he said, "so I was led to consider whether matter–antimatter symmetry might get violated in black holes. During the period I talked to Treiman and Zee daily, and they quickly joined me in exploring the ideas. We soon realized what conditions would be needed for matter–antimatter asymmetry to occur."

Wilczek said he wanted to avoid grand unified theory in developing the calculations. "At the time I thought those theories were terribly farfetched," he said. But he eventually realized they had to be used.

I asked Wilczek about his first reaction to Yoshimura's paper. "His paper arrived as we were thinking about these things," he said. "At first I thought very little of it because of the mistakes that were in it, and because it relied on grand unified theory. But once we understood things clearly the true value of the paper was more evident to me; indeed he had most of the essential ideas before we did."

Shortly after Wilczek and his group published their paper, another was published on the same subject by Steven Weinberg. Weinberg, as I mentioned earlier, was one of the inventors of the electroweak theory. Wilczek talked about Weinberg's paper. "For a couple of weeks I was on the telephone almost daily with Weinberg," he said. "These conversations were very helpful in clearing up the concepts. The letter he wrote (with a strong acknowledgment) was much more straightforward than our paper and really made things accessible and popular to the larger physics community."

SUMMARY OF THE DEVELOPMENT OF ASYMMETRY

Because of the work of Sakharov and the others mentioned above, scientists now feel they have a good idea why our universe, despite starting with equal amounts of matter and antimatter, now contains only matter. Their ideas are not yet set in marble; serious problems remain and the theory may eventually turn out to be totally incorrect. Nevertheless, cosmologists are happy that we have, in a logical and consistent way, been able to explain the asymmetry. The major problem is that the explanation depends on unified field theory, which in turn depends on the decay of the proton. And, of course, so far we haven't caught a proton in the act of decaying. Still, things look encouraging.

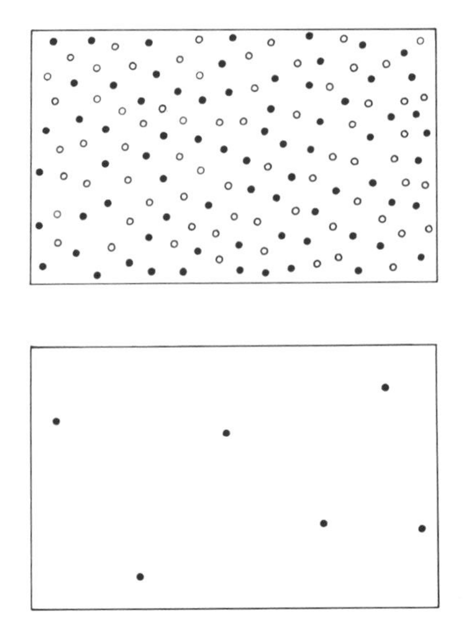

Lower: Universe before annihilation. Upper: Universe after annihilation. ●, *particle;* ○, *antiparticle.*

The key is the nonconservation of CP, along with violation of baryon number (and a change in the equilibrium state of the universe). In the very early universe, before about 10^{-35} second the temperature was on the order of 10^{28} K. This is higher than the mass of the X particle, and therefore X particles should have existed in large numbers. They would, of course, have been moving back and forth between the quarks, leptons and their antiparticles, changing one into the other. Furthermore there would have been an equal number of anti X particles (we will designate them as $\bar{X}$).

It is now well known that X's can decay in two different ways: either into a pair of quarks, or into a lepton and an antiquark. In practice large numbers of both of these pairs would be produced in the decay of a single X. The $\bar{X}$, on the other hand, can decay into a pair of antiquarks or an antilepton and a quark.

And if CP were conserved these decays would produce equal numbers of particles and antiparticles. But it is not. And because of this, slightly more particles were produced than antiparticles—only about one more for every billion particles and antiparticles. But this was enough.

As the universe cooled below the threshold temperature for the production of X particles the excess was frozen in: one billion and one particles for every billion antiparticles. Then came annihilation, and in time each of the particles found an antiparticle and annihilated, creating photons. By the time the universe was a few minutes old there was nothing left but the small residue of matter. Indeed, this is what the entire universe—the stars, galaxies, and even you—consists of today.

Because of this gigantic annihilation the universe should now consist mostly of photons. Does it? Indeed it does. With the discovery of the cosmic background radiation we realized that there are roughly a billion photons to every baryon in the universe. And we feel relatively confident that we know why.

CHAPTER 9

The Cosmic Cookbook

The asymmetry problem appeared to be solved. After matter and antimatter annihilated, only matter was left. The universe then consisted of electrons, neutrinos, photons, and a few protons and neutrons. But we know that it now contains many complex atoms—heavy elements. Where did they come from? How did they form? Gamow, as we saw earlier, was convinced that they came about as a result of collisions involving neutrons. He believed a step-by-step process occurred in the early stages of the big bang that produced all the atoms. But he was quickly stopped after helium-4 by a gap that was impossible to jump (and also at atomic mass 8).

In this chapter we will take a closer look at this problem. The first thing we have to ask ourselves is: If the elements were not formed in the early universe, where were they formed? The only reasonable alternative is the interior of stars. The temperatures in the cores of stars are not as high as they were in the early universe, but as we will see, stars have certain advantages.

Despite the difficulties, not everyone had given up on the early universe. Gamow's two students, Ralph Alpher and Robert Herman, continued to work on the problem. In 1953 they teamed up with James Follin to take a closer look at the problem. They soon realized that they would have to deal not only with neutrons, but also with protons—both would exist in the early universe. Taking both into consideration they traced the sequence of events as the universe cooled. Starting about 10^{-4}

second after the big bang they managed to give a fairly accurate description of everything that occurred up to a time of about 30 minutes.

But the gaps remained. They could find no way to cross them. Their paper, "Physical Conditions in the Initial Stages of the Expanding Universe," was published in 1953. But by then attention had switched away from the early universe. Although temperatures in the interiors of stars were much lower than those in the very early universe, stars had an advantage—an important one, as it turned out. Their core density was high, and therefore collisions of particles were frequent.

ACCORDING TO HOYLE

Fred Hoyle of Cambridge University was one of the first to take a serious look at stars as the birthplace of the elements. Born in England in 1915, Hoyle attended Cambridge, staying on after graduation to teach, becoming Plumian professor in 1958. Although he is best known for his steady state theory of the universe (for several years the major competitor to the big bang theory), which he formulated with Gold and Bondi in 1948, Hoyle has made many important contributions to astrophysics and cosmology. Furthermore, he was the founder of the Institute of Theoretical Astrophysics at Cambridge, and for a number of years served as its director. In recognition of his many contributions to science he was elected fellow of the Royal Society and honored by knighthood. And, despite a busy scientific career, he somehow found time for several other activities: he has written several excellent science fiction novels, one of which was made into a planetarium program, several popular science books, and even a musical comedy that was produced in London. All in all quite an accomplishment for one person.

One of Hoyle's major contributions was helping to determine how elements were formed in stars. Some people have said that he did this only because he knew they could not be

Fred Hoyle. (Courtesy AIP Niels Bohr Library, E. E. Salpeter Collection.)

formed in his steady state universe. (The steady state theory, which assumes that the universe has always been the same, and will continue to be the same into the infinite future, is no longer accepted.) But this is not true. Hoyle had been thinking of the problem even before World War II was over. The steady state theory was not published until 1948.

Toward the end of the war Hoyle made a flight from England to San Diego in conjunction with his work on radar. Realizing that he was close to the observatory at Mt. Wilson he decided to visit it. The place was almost deserted, most of the astronomers having been assigned to war-related projects. But one of them, Walter Baade, a recent émigré from Germany, was not trusted with classified work, and therefore left at the observatory. Hoyle talked to him, asking him about some of his ideas. Was it possible that the elements were "cooked" in stars? Baade was impressed with Hoyle's ideas and thought they should be pursued. Hoyle was encouraged. "After that talk," he said, "I began to think in a serious way about stellar interiors." It was the beginning of what was to be a fruitful trek.

But when he began to delve into the problem he found that

there were more roadblocks than he anticipated. In order to calculate the rates of the various nuclear reactions, he needed "cross sections." They are numbers that tell you how probable a reaction is, and therefore how likely it is to occur. And the only way you can get them is to measure them in the laboratory. Hoyle's initial search indicated that few such measurements had been made.

Then one day he was talking to nuclear physicist Otto Frisch, who casually mentioned that he had a table of cross sections in his office. Hoyle was overjoyed. "I was sure I could crank out the whole problem of nucleosynthesis [production of elements] in stars within six months," he said after looking at it. And he was soon hard at work. But still, things did not go as smoothly as he had hoped. The six months turned out to be a year, but finally in 1946 his first paper on the subject was published. It was titled, "On the Synthesis of Elements from Hydrogen," and although it was far from complete, it laid the groundwork. Still, at this stage, Hoyle could not yet explain the production of an element as simple as carbon. Then in 1949 he began thinking about the possibility of three particles coming together simultaneously—three helium nuclei, or alpha particles as they are commonly called. He gave the project to a graduate student as a thesis project. Everything was going well, when to Hoyle's dismay, the student disappeared. Hoyle was particularly disappointed because the student was about two-thirds of the way through the project. "Whether he left because he didn't like me, Cambridge, the problem, or himself, I don't know," said Hoyle. "Anyway, he left." Hoyle thought about continuing with the problem on his own, but he had just finished a tussle with several referees over the publication of his steady state theory, and was in a depressed mood.

He looked over the work that the graduate student had done, and filed it away. Then one day in 1952 he was going through a journal and saw an article by Ed Salpeter. Salpeter had used the triple-alpha process to show that carbon could be produced in stars. Hoyle could not believe his eyes—the solution to the very same problem he had given the student a few years earlier. He cursed himself for being so stupid, and not completing the problem on his own.

Reading through the article, though, he saw that Salpeter's work had a flaw. Carbon would be lost as fast as it was produced. Nothing would remain. Hoyle was sure, nevertheless, that there was net production. There had to be. After studying the reactions for a while he saw that it was possible if there was a "resonant" energy state at 7.65 MeV. It was a crazy idea—but it had to be there. How could he find out if it was? Being at Caltech at the time, the logical place to go was Kellogg Laboratory. Willy Fowler and his group might know, or if not, they could easily check.

Hoyle walked into the lab and announced that there had to be an energy level near 7.7 MeV. It was needed so that carbon could be produced in red giant stars. Fowler was not impressed. He was familiar with Hoyle's work but had not taken it seriously. His first thought, as he said later, was, "Hoyle—go away and stop bothering us."

But Hoyle's recollection of the meeting was quite different. He was pleased because Fowler did not laugh at his crazy idea. In fact, to him, it appeared as if Fowler was interested in doing the experiment. And, according to Hoyle, there was soon a mob in the office, everyone talking about how the experiment could be done.

Within about a week it was performed—and alas—the state was found. Hoyle was right. And Fowler was astounded. "That made a believer out of me," he said. Until then most of those present at the meeting, and most astronomers in general, were sure that the elements were somehow produced in the early universe. Within a short time of the experiment, though, there was a dramatic turnabout. Hoyle had found a way to jump the gaps at 5 and 8. There seemed to be little doubt now: the elements were produced in red giant stars. Not only could a jump be made from helium to carbon, but it appeared that all the elements could be produced in stars. Of course, there was still the problem of determining exactly how. But that would soon come.

B²FH

Hoyle now had a convert—William Fowler, or "Willy" as he was called by his friends. And Willy would soon become a strong supporter and lifelong friend. Together they would write two of the most important papers in astrophysics. Born in 1911 in Pittsburgh, Fowler spent his early years in Lima, Ohio. Lima was a railroad town and it was there that Willy began a lifelong love for railroad locomotives. He attended high school in Lima, was on the football team, and worked as a recreational director during the summers.

Upon graduation he went to Ohio State University. He was still uncertain at this stage what he wanted to do, but he had won a prize for an essay on cement, so ceramic engineering seemed a logical choice. One of his first courses, though, was physics and he enjoyed it; the lab particularly impressed him—so much so that he soon fell in love with experimentation. Upon graduation he went to Caltech, eventually working under Charles Lauritsen at the Kellogg Radiation Lab. "Lauritsen was the great influence of my life," he wrote. "He taught me how to do physics, and how to enjoy it."

Fred Hoyle also had a strong influence on him. Fowler was, in fact, so impressed with Hoyle's prediction of a line at 7.7 MeV that the following year he took a sabbatical and went to Cambridge to work with him. Hoyle was, unfortunately, tied down with a heavy teaching load and previous research commitments, so the two did not get together often. But there were two other astronomers at Cambridge, Margaret and Geoffrey Burbidge, who had considerable time for research, and Fowler soon began working with them.

The following year Fowler, the Burbidges, and Hoyle got together again at Caltech. And, as we will see, it was perhaps a

William Fowler.

lucky accident that they did. Interestingly, it almost never happened. The Burbidges only had a temporary position at Cambridge and were in need of something for the following year. Both were astronomers—Margaret an observer, and Geoffrey a theoretician. Fowler was enthusiastic about the collaboration, and felt that they worked well together. He would try to get them a position in the United States. He was sure he could get Geoffrey a temporary position at Kellogg Lab, but Margaret, being an observer, preferred access to a telescope. So Fowler wrote Ira Bowen, the director of nearby Mt. Wilson Observatory, to see if he could arrange a temporary position for her. Bowen wrote back that he could not offer her anything because there were no toilet facilities at the dome for women. When Fowler told Margaret she gave him a disgusted look and said, "I'll use the bushes." Fowler knew that Bowen would not appreciate the humor of this so he asked him if he could find a position for Geoffrey. In the meantime he arranged for a position for Margaret at Kellogg. Bowen finally came through with a position for Geoffrey, and as expected, each time Geoffrey went to observe, Margaret tagged along, so things worked out splendidly for everyone—and for the bushes.

Hoyle was at Caltech that year as a visiting professor. And soon the four of them were working out the details of element synthesis in stars. It was a monumental work requiring lengthy calculations. An early, short version of the work was published in *Science* in 1956. But the paper in its entirety ran to over 100 pages. Where would they publish such a long paper? Hoyle still had a bad taste in his mouth for referees. They were in the later stages of the work when Fowler bumped into an old friend, Ed Condon, who was now editor of *Reviews of Modern Physics*. He told Condon about the paper, and as it turned out, that was exactly the type of paper Condon was looking for: a review paper with important new results in it. "Send me the paper when you finish it and I'll publish it rapidly without having it refereed," said Condon. And they did, and indeed it was published without delay—much to their delight. "Those were the days," said Fowler later.

The paper eventually became known by the initials of its authors: B^2FH. It is now a paper that is known to all astronomers—one in which the production of the elements in stars was described in detail for the first time. Interestingly, a similar, but less complete, work was published by A. G. W. Cameron in the *Journal of the Astronomical Society of the Pacific* at about the same time. Fowler was awarded the Nobel Prize in 1983 for his part in this work and other contributions to physics.

The problem had been solved—the elements were made in stars. Hardly anyone gave the early universe a second thought after the publication of B^2FH.

But had it been solved? Within a short time observers began to measure the helium in the universe. And soon a puzzle began to arise. It is perhaps ironic that one of the first to notice it was Fred Hoyle—the very person who was determined to show that the elements were produced in stars, and the very person who was skeptical of any element production in the early universe. He was, in fact, sure there was no big bang—and therefore, no early universe. He soon realized that all the helium in the universe could not be produced in stars. There was too much of it—far too much. It was beginning to appear as if up to 25% of the material of the universe might be helium. In 1964 he, along with R. J. Taylor, published a paper titled "The Mystery of the Cosmic Helium Abundance." In it he went as far as suggesting that the universe may have began with a big bang, and the helium may have been produced shortly after this event. Strangely, though, he had not yet given up hope for his steady state theory. The admission that the big bang picture might be valid was obviously a hard one for him to make. But it seemed to be the only way around the problem. Still, aside from this, there was little evidence for a big bang.

Then came 1965. Penzias and Wilson discovered the cosmic background radiation, and its temperature agreed with the prediction made by Dicke and Peebles. Hoyle could hardly believe it. But he had to face the facts, and finally he began to take the big bang seriously. There had to be something to it. The abundance of helium in the universe pointed to it, and now there was the cosmic background radiation—radiation that was left over from the big bang.

BACK TO THE BIG BANG

With the discovery of the cosmic background radiation it was important that the early universe be reexamined to see what elements could be produced. In particular, it was important to calculate the expected abundances of the light elements in addition to that of helium-4 (which had been previously calculated approximately by others). Hoyle teamed up again with Fowler, both men eager to get involved in a new project. This time they were joined by Robert Wagoner, who had just graduated from Stanford. Was it possible, they asked themselves, that something had slipped by Gamow and his colleagues? After all, if Hoyle had not insisted that there was an energy level at 7.7

MeV, the route to element production via the stars might have evaded them. They had to check every possible route, every sequence of reactions through the light elements past helium. Wagoner took on the brunt of the work—the writing of a huge computer program.

Wagoner was born and raised in Teaneck, New Jersey. His first contact with physics was indirect; after reading several books on rockets in high school he decided to build one. But he soon found he was not cut out to be an experimentalist. "My chemistry laboratory course at Cornell was a disaster," he said. "I was destined to be a theorist." He said his physics teacher in high school was poor, but luckily he had an excellent mathematics teacher. At the time, though, he was leaning heavily toward engineering as a goal. And in the late 1950s he enrolled in the engineering program at Cornell. He majored in mechanical engineering, but wanted to switch to aeronautical engineering in graduate school. It was during his stay at Cornell that he attended a series of lectures on cosmology by Fred Hoyle. He was fascinated. "I read every book on cosmology I could get my hands on after hearing him," he said. "And it was quite a thrill actually working with him several years later."

After graduation, Wagoner moved on to a M.Sc. program in engineering at Stanford. "This program allowed me great freedom in my choice of courses, and it required no thesis," he said. "I therefore took advantage of it to make up my physics background." When he completed the program he applied to the physics department for permission to enter as a graduate student. He was worried that they would not accept him, but they did. And from 1962 to 1965 he worked in physics under Leonard Schiff. "It took a few years to change from thinking like an engineer to thinking like a physicist," he said.

After completing his Ph.D. at Stanford he moved on to Caltech. He was planning on working in relativity, but the microwave background had just been discovered and everyone was excited about it. Hoyle was there at the time. And Fowler asked Wagoner if he would like to work with him and Hoyle.

Robert Wagoner.

Wagoner quickly accepted. I asked Wagoner for some of his recollections of the project. "My role was to create the computer program and help Willy with the various nuclear reaction rates, looking up cross sections and so on," he said. "Fred provided the big picture, the cosmological overview, and he wrote the first draft of the paper. Willy provided access to all the cross-section data we needed and taught me how to use it." Regarding the two men he worked with, Wagoner said, "Fred had great insight. His ideas weren't always correct but they helped us ask the right questions. Willy kept us honest with the astounding breadth of his knowledge of nuclear physics. The original idea for the project came from him. He was quite familiar with Gamow's work and the later history . . . even more so than Fred, I think."

The computer program was a long one, and it was also a tricky one to run, according to Wagoner. "It would develop instabilities, and because of this it took over a year to get it to run to completion without producing nonsense." Reaction rates for almost 100 nuclear processes were included in the program, and many conceivable physical situations were considered.

Soon there was no doubt: nothing beyond helium could be produced in the big bang. Or, more exactly, "almost nothing," for there was a small amount of lithium produced. In short, then, deuterium, helium-3, helium-4, and lithium-7 were all produced in the big bang, and of them, helium-4 is the only one that is abundant. The program predicted that about 25% of the mass of ordinary matter in the universe should now be helium—in excellent agreement with observation. The other abundances also agreed with those observed. In 1967 they published their paper in *The Astrophysical Journal*. It was titled, "On the Synthesis of Elements at Very High Temperatures." And like B^2FH it was also a landmark paper.

This paper, coupled with the discovery of the background radiation, caused a resurgence of interest in the early universe. I should mention, incidentally, that about the same time Jim Peebles was doing similar work at Princeton. His paper was not as extensive as that of Wagoner, Fowler, and Hoyle, but nevertheless, it established basically the same results.

OBSERVATIONS

The major difficulty at this stage was that observation still lagged far behind. But it soon caught up. Helium has, in fact, now been observed both in stars and in galaxies. One of the major problems in relation to helium, though, is that although most of it was produced in the early universe, some was produced in stars (we will talk about the details of this in Chapter 14). The question is: What fraction was produced in stars? We now believe that it is less than 10%.

There have been many studies of the abundance of helium. In 1973 Robert Rood of the University of Virginia arrived at a percentage between 22 and 25 based on a study of helium in stars. In 1977 Demarque and McClure found a value close to 0.20, but not inconsistent with 0.25, in a study of clusters. And in 1978 Hirshfeld, in a study of dwarf galaxies, found a value of 24%.

These values are all consistent with predicted values. I asked Michael Turner of the University of Chicago what would happen if it was eventually proven that there was less than, say, 22%. "If the helium abundance were shown to be less than that, the big bang theory would be in serious trouble," he replied. "How would we get around the problem? One way is as follows. In order to get a lower limit to the helium that comes out of the big bang we need a lower limit on the number of baryons in the universe. In a paper we wrote in 1984 we tried to use tritium to lower this limit. We did some very detailed reasoning based on the fact that tritium is very difficult to destroy. But we couldn't make an ironclad case. A second possibility is that the universe was anisotropic at the time of nucleosynthesis. Some calculations indicate that this might decrease the amount of helium." He paused, then shook his head. "If we had to throw out the hot big bang theory . . . that would be sad, indeed."

But so far, observations of helium seem to confirm the big bang theory. Furthermore, observations of another element that is predicted to exist by the big bang theory, namely deuterium, strengthen the confirmation. Deuterium has, in fact, now been detected throughout the solar system—in the sun, on Jupiter, and even on the moon.

One of the first to search for it beyond the solar system was Sander Weinrab of the National Radio Astronomical Observatory in West Virginia. Using an 85-foot radio telescope he searched diligently, but found nothing. He finally came to the conclusion that its abundance had to be less than one part in 13,000.

But how would we detect deuterium? For an answer let us look at its structure. It is composed of a nucleus consisting of a proton and a neutron, around which whirls a single electron.

From a simple point of view we can think of both the electron and the nucleus as tiny spinning tops. The energy of the electron depends, in fact, on which direction it is spinning relative to the spin of the nucleus. It can spin in the same direction, or in the opposite direction. And occasionally—very occasionally— it may suddenly change its direction of spin, and when it does it emits radiation that has a wavelength of 92 centimeters. This means that if we can detect radiation of this wavelength with our radio telescopes we know that deuterium is present.

By 1970 several searches for deuterium had been made, and all were unsuccessful. Nevertheless, Diego Cesarsky and Alan Moffat of the Owen Valley Radio Observatory and Jay Pasachoff of Williams College decided to make another attempt. The best place to look, they decided, was in the direction of the center of our galaxy, which happens to be in the direction of the constellation Sagittarius. They elected to use the 130-foot radio telesope of the Owen Valley Radio Observatory. And finally in March of 1972 they were ready. There was, unfortunately, a difficulty. The center of our galaxy was visible above the horizon at Owen Valley for only about six hours a day. They decided to concentrate on it during this time and use any other time they had available to observe the Great Nebula of Orion and Cass A in the constellation Cassiopeia (a well-known source of radio waves).

The first runs looked promising. There was even a hint of a line at 92 centimeters. But there were a lot of other peaks (random fluctuations) near that wavelength and it did not show up clearly. They continued observing at two-week intervals. The faint glimmer of a line remained—but it was not sharp enough for them to be certain deuterium was really present. It could be a genuine line; on the other hand, it could also just be a random fluctuation. The following summer they tried again, and again the same barely visible line appeared at 92 centimeters. About the only conclusion they could draw was that the deuterium abundance was somewhere between one part in 3000, and one part in 30,000.

Then came the first positive detection. But it was not pure

deuterium. Keith Jefferts, Arno Penzias, and Robert Wilson observed deuterium cyanide in the Great Nebula of Orion. It was deuterium, but it was tied up in a molecule.

The first detection of pure deuterium came from observations using the Copernicus satellite. John Rogerson and Donald York showed that about 10% of the mass of the interstellar medium was deuterium. This has since been substantiated by several others.

Wagoner made use of most of the new results in 1973 when he published an updated version of element production. His earlier results had shown that the abundance of helium did not depend on the density of the universe at the time it was produced, but the abundance of deuterium did. This was borne out again in 1973. This meant that if we knew the abundance of deuterium at the time of light-element synthesis we could determine the present density of the universe. And when the calculations were made astronomers got a bit of a shock: there were not enough nucleons in the universe to close it.

A further update was made in 1977 when David Schramm and Jongmann Yang of the University of Chicago, Robert Rood of the University of Virginia, and Gary Steigman, who was then at Yale, used Wagoner's program to investigate the latest data. One of the important results of their paper was the constraints they were able to put on the number of different types of particles that could exist in the universe.

I talked to Steigman, who is now a professor of physics and astronomy at Ohio State University, about this work and his other work on nucleosynthesis. He has made several important contributions to the field. Steigman took his undergraduate work at the City College of New York. He said he had little interest in astronomy at the time. "When you live in New York City it's rare that you have an interest in the stars," he said. "You hardly ever get to see them. Besides, astronomy wasn't very exciting to me then . . . it was like stamp collecting . . . consisting mostly of classifying things. Particle physics was the exciting area. But when I went off to graduate school at Cornell

Gary Steigman.

important discoveries were being made: quasars, black holes and neutron stars. This was the first awakening I had that astrophysics was an exciting area. I left Cornell and came back to New York University, where my thesis advisor was Mal Ruderman. Ruderman had done work in particle physics, but also worked in astrophysics—on neutron stars and pulsars. And as a result I sort of oscillated back and forth between particle physics and astrophysics." Steigman said that for his thesis he did research on antimatter in the universe. This research involved astrophysics, nuclear physics, and atomic physics. "That was my first real particle physics–astrophysics connection," he said.

His involvement with nucleosynthesis came about as a result of his teaching of cosmology at Yale. In going through the section of nucleosynthesis for his course he realized that the helium abundance could place a constraint on the number of families of neutrinos (and therefore also the number of families of quarks). David Schramm and Jim Gunn came to the same conclusion at about the same time. Schramm and Gunn were at a workshop at Aspen where Steigman gave a talk on his discovery. "After the lecture I learned they had come to the same conclusion. So we joined forces and published a paper together," said Steigman.

In the paper that Steigman wrote with Schramm, Yang, and Rood, the latest results on the abundances of deuterium and helium were presented and compared to the predicted amounts. Good agreement was found. They looked into the number of neutrino types and concluded that at most there was only one undiscovered pair. And finally, they showed that according to the latest observations the universe could not be closed.

SUMMARY OF EVENTS DURING NUCLEOSYNTHESIS

As a result of the work that has been done on light element synthesis we now feel we have a good idea what occurred during the era of nucleosynthesis. The following is a brief account.

About one two-thousandths of a second after the big bang the universe consisted of photons, electron–positron pairs, neutrinos, and a few protons and neutrons. The temperature was about 10 K. In a very short time it would enter the radiation era, and photons would be the dominant particles of the universe. Although the protons and neutrons were small in number, they were important, for it was from them that the present elements of the universe would arise.

At this stage the particles were still in equilibrium; in other words they were being created and destroyed in equal numbers. But eventually, as the universe cooled the equilibrium would be broken. The number of neutrons at this point was equal to the number of protons, and this would remain so while the density and temperature were high. But neutrons decay into protons, with the emission of an antineutrino, and as the temperature dropped the number of protons became greater than the number of neutrons.

If things had continued in this way, eventually all the neutrons would have decayed. But this did not happen, because the temperature was soon low enough that when a neutron struck a proton it stuck to it, forming a nucleus of deuterium. And for a short time the universe was dominated by deuterium. But most of the deuterium nuclei were soon struck by another neutron, creating tritium. And tritium, being unstable, quickly decayed to helium-3. Fortunately, all the deuterium did not become tritium; a small fraction survived. And because this small fraction depended on the average density of the universe at the time, we now have a measure of what it was. At the same time, two deuterium nuclei were also colliding, forming helium-4.

Of importance during this era is the ratio of neutrons to protons. By measuring the present amount of helium in the universe we can, in fact, determine what this ratio was, because it depended on the amount of helium that was produced.

In a matter of minutes it was all over. Almost all the deuterium was converted to helium. And today we see about 25% of the material of the universe as helium. Thus, we can say with some confidence that hydrogen, helium, deuterium, tritium, and a small amount of lithium were formed in the big bang. But what about the other elements? Well, for those, Hoyle was right. They were formed in stars (and in supernova explosions). We will talk about them later.

It is perhaps interesting to ask why only the light elements were formed in the big bang. And why only the heavy elements were formed in stars. Of course we know that light-element synthesis was stopped by the gaps at 5 and 8. But aside from this why were they synthesized in this way? As we really do not know, I suppose the only thing we can say now is that the universe just happened to be put together this way. Perhaps some day we will find out why.

Anyway, the first nuclei were now present in the universe, and in about 10,000 years the universe would be cool enough for these nuclei to attract electrons and the first atoms would appear. When this happened an important event occurred. We will talk about it in the next chapter.

CHAPTER 10

Emergence of the Fireball

When the first atoms appeared in the universe the radiation decoupled from the matter and expanded freely into space. And as the universe expanded it cooled until today it has a temperature of only 3 K.

This cosmic background radiation, as it is now called, was discovered in 1965 by two Bell Laboratory radio astronomers, Arno Penzias and Robert Wilson. Like a number of other important discoveries, it was an accidental one. At the time Penzias and Wilson were converting a specially designed radio telescope for use in radio astronomy. They wanted to study the radiation emitted by the Milky Way galaxy. To do this they had to get the "noise" out of the telescope. After working on the problem for about a year they found to their dismay that regardless of what they did, a tiny "hiss" always remained. They became quite frustrated with it—to them it was a nuisance. Oddly enough, this "nuisance" eventually won them the Nobel Prize. As it turned out it was coming from deep space—it was an "echo" of the big bang explosion that created the universe.

Arno Penzias was born in 1933 to Jewish parents in Munich, West Germany—within weeks of Hitler's takeover. Although his father was born in Germany, he was a Polish citizen. So, besides being Jewish, Penzias was also Polish—the two groups that Hitler despised. Needless to say, his early life was chaotic. Penzias admits, though, that he never felt the frustration, turmoil, and the discrimination that was directed at his family, mostly because he was very young, and because he went to a Jewish school. He says, though, that he was a bit of a showoff when he was young, and was told by his parents to be careful of what he said in public.

Germany's prejudice against Poles grew until finally in 1938 it was decreed that all German Poles would be deported to their homeland. Poland, unfortunately, did not like Jews any more than the Germans did, and within a month of the decree they issued one of their own: after a specific date no Jews would be allowed to enter Poland. The Germans quickly rounded up all the Jews they could and rushed them to the Polish border, hoping to beat the deadline. But in the case of the Penziases they were too late. Arno and his family were forced to remain in Germany. But Hitler was not finished: he now gave all Jews still in Germany six months to leave. "The strain on my father and mother must have been tremendous," said Penzias. They knew that if they did not get out they would be sent to Dachau, a concentration camp. But as Penzias said, they somehow managed to keep all this from him.

The Penziases wanted to go to the United States, but had to have a relative willing to sign for them. And they had none in the United States. Fortunately, there were people who would claim they were related, and about a month before their deadline ran out somebody signed for them. But they also needed an exit visa. Such visas were easy to get for children, but extremely difficult for adults to get—particularly Jews. Their parents did not want to let them go alone, but finally, with the deadline only days away, they had no choice. So they put five-year-old Arno and his four-year-old brother on a train for England. They were sure they would never see them again. Fortunately, just before the deadline they were both able to get visas, and later joined their children in England. From there they sailed to the United States in late 1939. Shortly afterwards war broke out in Europe. "We made it [out of Germany] with very little time to spare," said Penzias.

His family settled in the Bronx, where Penzias went to elementary and junior high schools. "It wasn't a happy time," he said. His German accent clearly labeled him a foreigner; furthermore, because he was not good at sports he had a difficult time fitting in. High school was better. He went to Brooklyn Technical, where he had his first taste of physics. In February 1951, he enrolled at CUNY. He had taken a considerable amount of chemistry in high school, and sort of "drifted into chemistry in college." But a few weeks of it convinced him that he preferred physics. He says that although he did well in physics, he was not a top student and worried about what graduate school he would be able to get into. After a two-year stint in the army, however, he managed to get into Columbia, one of the top graduate schools in the country. At that time Columbia had several Nobel laureates on its faculty; among them was Charles Townes. Penzias took a class of his and because he did quite well in it he eventually got up enough nerve to ask him if he could work under him. Townes agreed and put him to work making a maser (an instrument for amplifying a microwave signal into a coherent beam). His thesis project consisted of using it in conjunction with a radio telescope to check hydrogen gas in galactic clusters. He admits he had a "dreadful" time with it and it was not a good thesis. One of the first things he hoped to do when he went to work was vindicate himself by doing it over properly. When he was nearing completion of his thesis he visited Bell Labs; several people were working on masers, and it seemed like a good place to continue his research so he applied for a job. In 1961 he was hired. Today he is vice president of research.

Robert Wilson, the codiscoverer of the cosmic background radiation, was born in 1936 in Houston, Texas. His father was a chemical engineer who worked in the oil fields. And, as might be expected, Wilson spent a considerable amount of time in the field with his father. Although he had a mild interest in the oil business, his real love was electronics. He was always tinkering with radios, and in high school made a considerable amount of pocket money repairing radios and televisions.

He went to high school in Houston. His favorite subjects were science and math, but he admits he was not an outstanding student—ranking only in the upper one-third. Things went much better, though, in college. In 1953 he enrolled at Rice, hoping to become an electrical engineer. But he soon found there were aspects of electrical engineering that he did not like, so he switched to physics, where he got straight A's.

His grades were so good that he had a choice between MIT and Caltech as a graduate school. He chose Caltech, enrolling in 1957. Caltech was one of the top schools in the nation, with Gell-Mann, Richard Feynman, H. P. Robertson, and Fred Hoyle among its faculty. While at Caltech, Wilson met John Bolton, a radio astronomer who had just come from Australia. Bolton had pioneered in radio astronomy and was at Caltech to form a radio astronomy group and to build a radio telescope. When Wilson joined the group, two radio telescopes had already been constructed in Owens Valley, about 200 miles north of Los Angeles. Working under Bolton, Wilson did a survey of the radio waves from the Milky Way at a wavelength of 31 centimeters. It formed the basis of his thesis. In 1962 he completed his degree, but remained at Caltech for a year as a postdoc. While there he was interviewed by a Bell Lab representative, and in the spring of 1963 he came to Crawford Hill. Penzias had already been there for two years.

BELL LABS AND THE HOLMDEL TELESCOPE

The radio telescope that Penzias and Wilson used in their discovery was built by Bell Lab engineer Arthur Crawford in the late 1950s for use in receiving reflected messages from Echo, a communications balloon. It was a strange-looking instrument, resembling a giant bugle, but it worked. Radar signals reflected via Echo from California were picked up by it. After its initial success with Echo it was left in place for use with Telstar, a Bell Labs satellite.

Robert Wilson and Arno Penzias. Holmdel telescope is in the background. (Courtesy Bell Laboratories.)

But by 1963 its usefulness in communications was generally over and Penzias and Wilson were given permission to convert it for use in radio astronomy. It was small compared to most radio telescopes, with a collecting area of only 25 square meters. But it was unique. It was extremely sensitive and could be used in a region of the electromagnetic spectrum (microwave) that had not yet been extensively studied. For large-diameter sources, according to Wilson, "it would be the world's most sensitive radio telescope."

To achieve this sensitivity it had to be used in conjunction with what is called a cold load, an artificial source cooled to liquid helium temperature that could be used as a reference. In this way they could make the telescope "noise free"—at least they thought they could.

In 1963 they converted the telescope to astronomical use,

and in 1964 were ready for their first measurements. They would look at the supernova remnant in the constellation Cassiopeia, called Cass A. Once this was completed they had several projects planned. Both men wanted to redo their Ph.D. theses. They selected a wavelength they thought would be particularly quiet—7 centimeters. But from the first measurement they had problems. "The antenna temperature was too high," said Wilson. "There was too much noise."

SOMETHING IS WRONG!

I should perhaps begin by explaining what this "temperature" is. What, in fact, do we mean when we say radiation has a certain temperature? First, it is important to note that all objects above 0 K emit radiation (photons). This means they have some "noise." 0 K, or absolute zero, is the lowest possible temperature in the universe. If you cool the interior of a black box to, say, 3 K, there will be a few photons in thermal equilibrium with the walls. By this I mean that the same number are absorbed as are emitted. If a radio telescope is placed in the box the message it receives will be a kind of "hiss" or static caused by the photons. As the temperature is increased the hissing becomes greater.

Penzias and Wilson found a hissing corresponding to about 3 K that they could not account for. They would have to determine where it was coming from before they could proceed with their program. They checked everything they could think of as possible sources. Manmade noise, particularly that emanating from nearby New York City, was checked. But when the noise in the direction of New York was compared to that in other directions there was little difference. They also examined the Milky Way as a possible source, but the radiation seemed to be coming from all directions—it was equal in intensity regardless of where they pointed the telescope—and this would not have been the case if it came from the Milky Way.

When they had completed the examination of all possible

sources it seemed as if the only thing left was the telescope itself. One possibility was the rivets used to construct the telescope. They were carefully covered with aluminum tape. Pigeons roosting in the throat of the horn were evicted and their mess cleaned up. But the "hiss" remained.

They had been working on the problem for several months when Penzias was talking to Bernard Burke of MIT on the telephone (January 1965). Burke asked him about his work with the Holmdel telescope. Penzias told him about the problem they were having. Burke, it turned out, had just heard about a talk given by Jim Peebles, a young theorist from Princeton. Peebles, who had just completed his Ph.D. under Robert Dicke, had argued that there should be some noise left over from the big bang explosion. "Why don't you phone Dicke at Princeton?" said Burke. "He might be able to help you with your problem."

Penzias called and told him about the problem. "That sounds very interesting," said Dicke. Dicke and Peebles had been investigating the properties of a pulsating, or oscillating universe and had become convinced that some of the radiation from the big bang should still be around. It would now have a very low temperature, but they believed it might be detectable. Peebles had, in fact, determined that its temperature should now be about 10 K (this was later reduced to 3 K). They had even convinced two Princeton experimenters, P. G. Roll and Dave Wilkinson, to look for it. Dicke sent a preprint of Peebles's paper to Penzias and Wilson; a few days later Dicke and Peebles, along with Roll and Wilkinson, visited Holmdel.

Dicke and his group were soon convinced that they were indeed detecting the leftover radiation from the big bang. Penzias and Wilson were pleased that there was an answer to their problem, but somehow they were not entirely convinced. Cosmology was still split at this stage, with some backing the big bang theory and others the steady state theory. Wilson, in fact, had taken a course from Fred Hoyle, creator of the steady state theory. He was therefore a relatively strong supporter of it. Still, any explanation was better than none, he said.

After the two groups discussed their results they decided to publish side-by-side papers in *The Astrophysical Journal*. The article by Penzias and Wilson was titled, "A Measurement of Excess Antenna Temperature at 4080 Megacycles/second." It was a short paper, only two pages long, in which they described the telescope and the efforts they had made to get rid of the excess noise.

"Arno and I were careful to exclude any discussion of the cosmological theory of the origin of the background radiation in our letter," said Wilson. They made reference to cosmology in only one sentence: "A possible explanation of the excess noise temperature is one given by Dicke, Peebles, Roll and Wilkinson in a companion letter in this issue."

The paper by Dicke and his group was titled, "Cosmic Black-Body Radiation." In it they talked about the expected residual radiation and the fact that Roll and Wilkinson had set up a detector in an effort to detect it, but had not yet been able to. They then mentioned the Penzias–Wilson discovery.

I asked Jim Peebles what his reaction was when he heard about Penzias and Wilson's discovery. He said his first reaction was relief. They had done a considerable amount of theoretical work in predicting the fireball radiation, and this seemed to be a verification. At the same time, however, he said he was somewhat "dismayed because he had let the cat out of the bag." Roll and Wilkinson had begun to search for the radiation, and it seemed that they would easily make the discovery before anyone else. But apparently they had not. And Peebles was a little embarrassed about it. Finally he also had some doubt. "There were so many other possible explanations of the results," he said.

Most scientists were excited about the news—but not all. George Gamow was furious when he saw the prediction. He shot off a letter to Dicke informing him that he and his students Ralph Alpher and Robert Herman had already made the prediction years earlier. Gamow pointed to a 1953 article of his in *Proceedings of the Royal Danish Society* in which he had made a prediction of 7 K. But a glance at the paper shows that it makes little sense: the argument he uses is circular (you need to know the temperature to calculate it). His students Alpher and Herman, however, did make a prediction of 5 K in 1948. They also went to the trouble of asking several groups of radio astronomers if it was possible to measure it. They were told no. Wilson believes, however, that it would have been possible at that time to make the measurement using World War II receivers.

It is also interesting that two Russians, A. G. Doroshkevich and I. D. Novikov, published a paper in 1964 pointing out that the radiation should be detectable. They even pinpointed the Holmdel telescope as being the most appropriate instrument for the measurement. Unfortunately, they erroneously concluded from another Bell Lab paper by E. A. Ohm that Gamow's prediction had to be ruled out.

Of further interest is the indirect detection of the radiation 30 years earlier by an entirely different means. Walter Adams of Mt. Wilson was studying spectral lines in interstellar gas when he discovered several faint lines. Among them were lines of cyanogen. Andrew McKellar of the Dominion Observatory in Canada, upon analyzing his data, discovered that they indicated the interstellar clouds had a temperature of about 2.3 K. This result later appeared in a well-known book on spectroscopy. But nobody apparently thought of it as the temperature of space.

Why was the radiation not discovered earlier, you might ask? Weinberg, in his book *The First Three Minutes*, discusses this question at length. One of the problems, he points out, is that there was not a lot of interest in the 1948 Alpher–Herman prediction and it got little attention. Dicke's group had not heard of it, and neither had Penzias and Wilson. Furthermore, in a subsequent paper by Alpher, Herman, and Follin in 1953 the problem was left for "future studies." Weinberg points out that it was a "classic breakdown in communications between theorists and experimentalists." Alpher and Herman, as I mentioned earlier, did talk to experimentalists about the measurement but failed to get anyone interested. This may have been because most scientists did not take cosmology, and particularly the early universe, seriously at that time. As Weinberg writes, "It is always hard to realize that the numbers and equations we play with at our desks have something to do with the real world." He also says that because of this we sometimes do not take our theories seriously enough.

Jeremy Bernstein and Gerald Feinberg, in their book *Cosmological Constants,* point out that part of the problem was Gamow's personality. "He was a larger-than-life Russian eccentric with an irrepressible and irreverent sense of humor. It is hard to know how seriously he himself took some of his scientific work, as brilliant as it often was," they write.

ORIGIN OF THE BACKGROUND RADIATION

Let us turn now to the background radiation itself. Where did it come from? Why is it all around us? To answer these questions it is best to begin witn a brief review of some of the events of the early universe. About a hundredth of a second after the big bang the temperature was still exceedingly high—about 100 billion degrees. The universe at this stage consisted mainly of electrons, positrons, neutrinos, and photons, but thinly distributed through them were protons and neutrons. The temperature was still much too high for nuclei to exist. Finally, though, at about 3 minutes the first nuclei appeared. But several hundred thousand years would pass before these nuclei collected electrons and became atoms. Until this time the photons were trapped—absorbed or scattered almost immediately after they were emitted. They could not escape the influence of the matter. They were in "thermal equilibrium" with it.

Strangely, though, despite the fact that there was considerable light in the universe, it was opaque. If you were there, you would seem to be in a thick fog. Finally, though, at a temperature of about 3000 K the nuclei began capturing electrons, and as they did the photons broke loose and flooded out into the universe, filling all space uniformly. And, almost like a miracle, space became transparent.

Looking out into the universe today we see these photons coming to us from all directions. Their intensity is the same in all directions; in other words, they are isotropic. What we are seeing is photons from the "shell" of last scattering. This shell still surrounds us, but is moving away at nearly the speed of light. As we look outward at it, we are looking back in time. It is the most distant thing we see in the universe—far beyond the most distant galaxies. It is, indeed, the "echo" of the big bang.

Penzias and Wilson had detected radiation that appeared to be coming from all directions. But was it definitely coming from the big bang? How could we be certain? It turned out there was a way. The intensity of the radiation from the big bang would have varied with wavelength in a particular way. If you made a plot of the intensity of this radiation against wavelength, you would get what is called a blackbody curve. (A blackbody is an

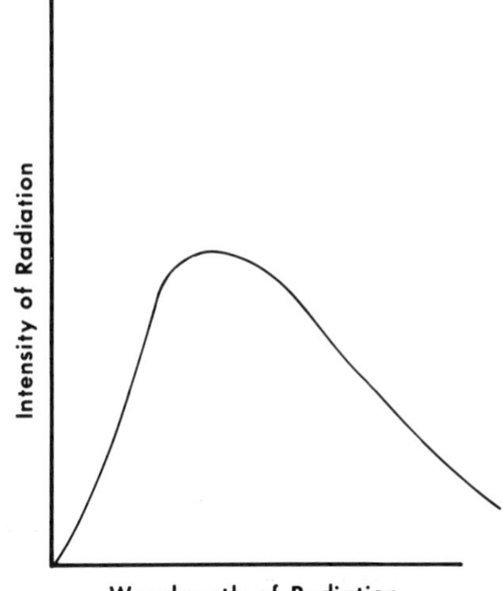

Blackbody curve.

object that absorbs and emits all radiation that falls on it.) We can easily construct such a curve. We need merely heat a blackbody and record how many photons are emitted at various wavelengths. The figure shows its general overall shape. Looking at it we see that, starting at long wavelengths and moving to shorter ones, we find the intensity increases. Finally, though, it reaches a peak (dependent on the temperature) and falls off rapidly.

Did the radiation that Penzias and Wilson detected satisfy such a curve? It was important to find out. Penzias and Wilson's point was at 7 centimeters; shortly thereafter Roll and Wilkinson got another at 3 centimeters. It was on the curve. Several other groups soon became interested and other points were obtained. All appeared to fit the curve reasonably well. There seemed to be little doubt that it was radiation from the big bang.

But there was a problem: all of the points were on the same side of the hump. It was important that they get points on the other side of it to verify that it indeed did fall off. The trouble with this is that our atmosphere interfered in this region and we had to get above it to get the required data. Balloons and rockets were needed.

The first points that were obtained using rockets created a shock among the scientists involved in the experiment. They were well above the curve. Maybe this was not the cosmic background radiation after all. But then it was discovered that the detector had picked up a small amount of heat from the rocket exhaust. When corrections were made the points came down to the curve—much to the relief of the scientists.

Then in 1975 D. P. Woody and several colleagues from the University of California and Berkeley Labs used a liquid helium-cooled balloon to accurately measure this region. They got a fairly good fit—but again there were small discrepancies.

THE SEARCH FOR ANISOTROPY

When Penzias and Wilson checked the background radiation they found that it was the same in all directions. In other words, it was isotropic. But their measurements were only accurate to about one part in ten. Was it possible that there was anisotropy that would be apparent with more accurate measurements?

To see the importance of a possible anisotropy let us go back to the late 19th century. Light had just been shown to have wave properties. But if it was a wave, some sort of medium was needed to propagate it, just as a wave on a lake needs water to propagate it. Scientists therefore assumed there was a propagating substance throughout the universe; they called it the aether. This aether was a mysterious substance; it had to be invisible, yet extremely rigid.

There was, however, a problem with the idea. The aether would act as a reference system for the universe. To see what I mean by this, assume you are sitting in a boat on a very calm lake (the lake is the reference system in this case). You wonder if you are moving. How do you find out? The best way is to set out a small buoy and see if it moves away from you.

If we wanted to find out if we were moving relative to the aether we would, in a sense, also have to set out a buoy. In 1887 Albert Michelson and Edward Morley set out to do this using an instrument called an interferometer that Michelson had just invented. But, to their surprise, they discovered that we were not moving relative to the aether. And a few years later Einstein showed that the aether was not needed—indeed, it did not exist. The wave aspect of light was included in the particles that made up light, namely, the photons.

With no aether around us we could not determine our velocity relative to the rest of the universe. But then came the discovery of the cosmic background radiation. And it was similar to the original hypothetical aether: it was uniform and filled the entire universe. We could therefore perform a "new aether experiment," and it would tell us how fast we were moving relative to the background radiation.

How would we go about such an experiment? To see, suppose we are in a thick uniform fog and cannot see a thing except the fog, but we need to know how fast we are traveling through it. We could find out by measuring its opacity (thickness) in all directions around us. It would appear thickest in the direction that we were moving. In the same way, if we had radiation at a temperature of 3 K (or more exactly, 2.7 K) all around us and we were moving in a particular direction through it, it would appear to be slightly hotter in this direction. Correspondingly, it would be slightly cooler in the opposite direction, with a gradual change between these two extremes.

In the mid-1960s Dave Wilkinson and Bruce Partridge set out to see if this was the case. Their apparatus had an accuracy of about one part in a thousand. But even with this accuracy it still appeared isotropic. More accurate experiments were soon set up, however. The first came in 1969; it was conducted by Edward Conklin. The second came in 1971 and was under the direction of Paul Henry. Although both hinted at anisotropy, there was still considerable uncertainty. But emissions from water vapor and other things were now becoming a problem. To get better measurements experiments would have to be done above our atmosphere.

Two groups took up the challenge. One consisted of Richard Muller, George Smurf, and Mark Gerenstein of the University of California at Berkeley. They enlisted the services of NASA and managed to get access to a U2 spy plane that was capable of flying at tremendous altitudes. The second group consisted of Dave Wilkinson and Bruce Corey of Princeton University.

Muller and his group got the first measurements. Their initial flight took place in July 1976. There was no doubt this time: there was anisotropy. The temperature of the radiation was highest in the direction of the constellation Leo, and lowest in the opposite direction in the sky. And there was a smooth variation between the two regions. Soon afterward, Wilkinson and Corey obtained a similar result.

What is the significance of these measurements? They meant that our solar system is moving in the direction of the constellation Leo at a speed of about 600 kilometers per second.

In fact it is not just our solar system, but our entire galaxy and several nearby galaxies that are moving in this direction. A new and important tool was obviously available and astronomers were quick to exploit it.

RECENT EXPERIMENTS

One of the major reasons many cosmologists were excited about the discovery of the cosmic radiation is that it gave them a tool to examine the universe at the time the cosmic background radiation broke free from the matter. This was when it had a temperature of about 3000 K, which was before there were any astronomical objects in the universe. But we know galaxies did eventually form so that there must have been small fluctuations in the material of the universe at this time. Looking at the radiation we should still be able to see these fluctuations. Can we? It turns out that so far we have not, even though we can now measure the radiation to one part in about 10,000.

J. Uson of Princeton University, using the 140-foot radio telescope at Green Bank, has shown that the background radiation appears to be perfectly isotropic. Incidentally, I should mention that the anisotropy that I talked about earlier is only an apparent anisotropy, caused by our motion through it. Uson has shown there is no actual anisotropy. And others have come to a similar conclusion.

Why is this a problem? The major reason is that, according to the standard big bang model, the universe is not causally connected. We talked about this in Chapter 7 in relation to the horizon problem. Yet the background radiation appears to be perfectly uniform. Of course, as we saw earlier, inflation theory—assuming it is correct—gets around this problem.

In the last few chapters we have looked at how the universe evolved once it came into being, but we have said nothing about where it came from in the first place. Now that we have a little background let us turn back and see if we can answer this.

CHAPTER 11

Before the Big Bang

What caused the big bang? What was here before it? These are, of course, not easy questions to answer. Indeed, before the last few years most cosmologists would not have dreamed of tackling them. They seemed outside the realm of science—a topic belonging to religion, or metaphysics. This point of view has recently changed, however. Scientists are now seriously considering such questions—and with some success.

There is a major difficulty in trying to go back in time to the instant the big bang occurred. Once we are inside the Planck era we no longer have a theory that describes the universe. General relativity, the theory that we usually rely on, does not work here. Space and time are just too chaotic at this stage for any known theory to describe it.

Despite this, several physicists have speculated on what this era would be like. We talked about it earlier. According to some it would be like a "space-time foam," a disconnected bubbling froth of tiny black holes popping in and out of existence. We also believe that only one type of particle existed—a superparticle. And correspondingly, there would have been only one force. In fact, it would have been impossible to distinguish between matter particles and force particles. To further complicate things the universe may have had many more dimensions than it has today. Our universe now has three dimensions of space and one of time; it may have had a total of 10 or 11 dimensions then.

One of the first questions that comes to mind when we begin looking at the details of the big bang is: What was here before it occurred? Was there a sea of empty space? Or was there just "nothing"? The question is, without a doubt, a difficult one to answer, but it is one that scientists are beginning to take seriously.

"MAYBE IT WAS A QUANTUM FLUCTUATION"

One of the fundamental ideas now being considered in relation to the origin of the universe came from Edward Tryon, now of Hunter College, New York. Born in 1940 in Terre Haute, Indiana, on September 4, 1940, Tryon became interested in science at a young age. He took his first class in physics his junior year in high school, and immediately fell in love with the subject. "Within a week of starting the course," he said, "I knew I was going to become a physicist." He remembers being told by the teacher that there was no end to the universe—it went on forever. The statement had a profound effect on him. He tried to visualize what it meant, trying in his imagination to get to the end of the universe. But he could not. "I felt a new door had been opened to me," he said. "It was a heady experience."

In 1958 he went to Cornell University with questions about the universe still spinning in his head. He read as many books on cosmology as he could get his hands on. But by his senior year he had found another fascination: quantum mechanics. And he soon became mesmerized by the philosophical implications of the theory. On the last morning of lectures Tryon stayed after class to take a picture of his professor, Nobel Laureate Hans Bethe, when, to his surprise, Bethe invited him to lunch. "I was delighted," said Tryon. "I had so many questions I wanted to ask him." Over lunch Tryon told him about his struggle to find new significance in the quantum mechanical wave function, and asked him for his views. After a few moments' silence Bethe replied, "Our intuition is based on our experiences in the

Ed Tryon.

macroscopic world. There is no reason to expect our intuition to be valid for microscopic phenomena." The statement had a profound influence on Tryon. He realized that if he were to do anything of fundamental significance in physics, he would have to set aside prejudices about the physical world he had built up over a lifetime.

In 1962 he completed his Bachelor's degree at Cornell, where he had minored in philosophy while majoring in physics, and began graduate work at the University of California at Berkeley. And just as Bethe had had a strong influence on him at Cornell, another soon-to-be Nobel laureate, Steven Weinberg, would have a strong influence on him at Berkeley.

When Tryon arrived at Berkeley he knew nothing about Weinberg, who was then only 29 years old. But a talk given by Weinberg on the importance of elementary particles in astrophysics impressed him. He saw for the first time that there were links between the very small and the very large. "The potential of the subject excited me," said Tryon.

Tryon took courses in quantum field theory and general relativity from Weinberg and when the time came for him to begin a thesis he realized that, notwithstanding the size and eminence of Berkeley's faculty, there was only one person he wanted to work under, and that was Weinberg. It was obvious that Weinberg was extremely busy, however, and might not be interested in taking on another student. After much hesitation Tryon got up enough nerve to ask, and was delighted when Weinberg said he would consider him. But first he would have to complete a test project. Tryon solved the trial problem in greater detail than even Weinberg had had in mind, and he was taken on. For his thesis, Tryon chose to study the relationship between general relativity and quantum field theory. In these subjects lay the seeds for Tryon's later conjecture about the origin of the universe: We have seen that general relativity describes the structure of the cosmos, and quantum field theory describes the spontaneous creation of particle pairs in a vacuum.

The cosmological problems that he had first considered many years before were still on his mind. Mach's principle, in particular, fascinated him. He thought about it off and on in the moments he was not working on his thesis. Late one afternoon, while pondering it, he began playing with the pertinent equations, and soon found that he could derive an expression for the universal gravitation constant (G) in terms of the total mass of the universe. He substituted numbers into the expression he had derived and found to his amazement that it gave G to an accuracy of 3%. He was ecstatic, sure he had made a discovery of profound significance. He rushed out of his room looking for someone to share his excitement with and soon found a professor he knew. The professor was amused by Tryon's exuberance, but, after looking at the derivation, he said with a shrug, "It could just be a coincidence."

Early the next morning Tryon went to see Weinberg—but was soon crestfallen. Weinberg explained that it was a known relation; Dicke and Brans had used it in a theory of gravitation they had devised, but their theory was not borne out by observations.

Tryon was not ready to give up, however. The relationship seemed so fundamental; there had to be a deeper meaning to it. He kept looking it over, feeling sure there was a profound message within it. But nothing came to him.

Upon completion of his thesis he went to Columbia University on a postdoctoral. His research changed focus to the strong interaction between subnuclear particles, but the derivation kept nagging at him.

One day in 1970 Dennis Sciama of Cambridge University was giving a talk on cosmology at Columbia. Tryon was unable to follow the details, and his mind wandered. At one point Sciama paused for a few moments to collect his thoughts and Tryon blurted out, "Maybe the universe is a vacuum fluctuation!" Everyone around him, including three Nobel laureates who were in the audience, began to laugh, thinking he was joking. Tryon, who was a junior member of the department at the time, was far too embarrassed to admit that he had made the suggestion seriously, so he kept quiet. In fact, his embarrassment was so great that he forgot the idea and completely blocked the incident out of his mind. It was not until several years later, after he had rediscovered and published the idea, that a member of the audience reminded him of the incident and he was able to recall it.

In 1971 Tryon left Columbia and went to Hunter College in New York City. A few months later a former colleague wrote, asking him if he would be interested in writing a chapter on cosmology for a book surveying modern science. Tryon replied that he would be pleased to. He then spent several months preparing for the work, reading everything he could get his hands on related to cosmology. As part of his research he went through every issue of *Nature* for several years back, looking over every article on cosmology that had been written.

He was organizing his thoughts one afternoon early in

1973, trying to assimilate everything he had read, when he began to think again about the "cosmic coincidence" he had discovered several years earlier. He began to think of it in light of what he had learned about cosmology since that time. The derivation, when he had made it, had seemed like a straightforward derivation of G, the gravitational constant. That interpretation had been pursued by Dicke and Brans, however, and seemed a dead end. Tryon asked himself once again what deep message lay in the relationship between G and the mass of the universe.

Suddenly and unexpectedly it came to him. In his mind's eye a brilliant burst of light appeared, expanding rapidly outward. In the same moment he knew that he was witnessing the creation of the universe and that he understood how and why it had happened. He was filled with awe at the simplicity and beauty of the creation process. He also sensed this revelation to be a unique moment in a quest that had spanned many ages—he pictured a group of cavemen gazing at the night sky, wondering where it had all come from, and he felt a kinship with them.

"I had been looking at the problem in the wrong way," he said. Turning things around, Tryon saw something quite different coming out of the calculation: the gravitational potential energy of every piece of matter in the universe was equal and opposite to its mass energy. The universe had a net energy of zero—and therefore it could be created from nothing. Not only *could* be, but *would* be: quantum theory tells us that any event that is not forbidden will occur, sooner or later, with the timing governed by laws of probability. Given that zero net energy is possible for a universe, quantum field theory implies that such universes will be inevitably created, spontaneously, via pair production in the vacuum. "I was momentarily stunned when I realized this," he said. The feeling was the most extraordinary one he has ever experienced.

Let's look at these ideas more carefully. The mass energy of the particles in the universe is their energy according to Einstein's famous equation that relates energy to mass and the speed of light. This, according to Tryon, appeared to be balanced by their mutual potential energy due to gravitation. We know that any two objects in space are attracted to one another with a force given by Newton's law of gravity. If such objects are initially stationary and are left to themselves, they will begin to fall toward one another. Since they gain energy of motion, or kinetic energy, as they fall, energy conservation implies that they had *potential* energy associated with their initial positions. As they fall toward one another, this potential energy is converted into kinetic energy.

It is customary (though not strictly necessary) to say that objects have zero potential energy when they are infinitely far apart. Since the potential energy decreases as objects fall toward one another, it must become negative for any finite separation. The relation between G and the mass of the universe that Tryon had discovered is consistent with the view that the negative potential energy of all the particles in the universe precisely cancels their positive mass energy, for a net energy of zero. Turning things around, this balance is *explained* by the idea that the universe was created from nothing.

Tryon had realized that if the universe had a net energy of zero, the concept of quantum fluctuations (pair production in the vacuum) could explain the creation of the universe. Let's take a moment to consider these quantum fluctuations. We talked about them earlier, but it is perhaps useful to remind ourselves what they are. It is well known that space, although it seems to be empty, is actually filled with virtual particle pairs—pairs that are created spontaneously out of the vacuum. Pair production violates energy conservation, but temporary violations occur in quantum theory: the greater the violation, however, the briefer the time it can last. An electron–positron pair, for example, has a mass energy of just over one MeV, and lives for a mere 10^{-21} second.

When a small number of pairs is created, the potential energy among them is negligible. The potential energy is proportional to the square of the number of particles, however, while the mass energy is proportional to the number itself. If by chance a great many particles were created close together almost simultaneously, their negative potential energy could balance their mass for a net energy of zero, and the particles could last forever. This was the kind of quantum fluctuation Tryon was thinking of: the spontaneous creation of an enormous number of particles whose net energy was zero, and which have survived to the present as our universe.

Tryon was soon hard at work writing an article for *Physical Review Letters* outlining his ideas. After submitting it he waited anxiously for a reply. As in the case of all articles it was sent to two referees for their comments. Finally the reply came: both referees recommended rejection. Tryon was stunned. Reading their criticisms he saw that neither had said the idea was wrong, but both were skeptical. And both felt that the article required much more work before it could be published. They suggested it should be "fleshed out" with mathematical details and sent to *Physical Review*.

Tryon did not know what to do. Working out the mathematical details might take years. And besides, he was uncertain how to proceed. Literally nothing had been done before to suggest how a detailed calculation might be carried out. Nevertheless, he was convinced the idea should be brought to a wide audience. Perhaps someone else would be interested in it and see how to develop it.

While he had been reviewing material for his chapter on cosmology he had become acquainted with the British journal, *Nature*. He remembered that they occasionally published articles that were rather speculative. Maybe they would be receptive. He therefore wrote up his idea as a "letter to *Nature*," hoping they would find it interesting enough to publish in this unassuming format. To Tryon's surprise and delight, the editor was fascinated. He liked the idea so much, in fact, that he decided to publish it as a regular article, rather than just a letter. It appeared in December 1973 with the title, "Is the Universe a Vacuum Fluctuation?"

"It generated considerable response," said Tryon. "I received over 140 requests for reprints." Articles on the idea appeared in *Science News*, and in several other scientific magazines. The BBC interviewed him, and broadcast the interview over their worldwide network.

But nothing further happened. Interest soon waned, and other scientists did not develop the idea as Tryon had hoped they would. Not until 1978, five years later, did another article on the subject appear. It was published in the *Annals of Physics* by R. Brout, P. Englert, and E. Gunzig of the University of Brussels. Their paper, unlike Tryon's, was "fleshed out" with mathematics. They proposed that the initial fluctuation occurred in a flat, four-dimensional spacetime, and that the initial pair, which was presumably superheavy, stimulated the creation of another pair which in turn stimulated others. Eventually the space became severely curved, which caused the process to get out of control, creating eventually all the particles of the universe.

Their idea was intriguing, but still there was little response from other physicists. The theory remained highly speculative, and the mathematics used by Brout and his coauthors was difficult to follow. Furthermore, if the universe did begin as a quantum fluctuation, it had to begin with equal numbers of particles and antiparticles. And this symmetry would likely have remained. But evidence was now coming in that this was not the case. The universe appeared to be made up entirely of particles (matter).

And there were other problems. Why was the universe so large and long-lived? Quantum fluctuations such as the one visualized by Tryon seemed likely to produce a much smaller universe than ours. Also, why was the early universe hot? This was also not explained.

Oddly enough, one by one, the problems were overcome. First, it was shown that even if our universe began with equal amounts of matter and antimatter it could evolve so that only matter remained, if CP invariance and baryon number conservation were violated, as happens in grand unified theories. We talked about this in Chapter 8. Then astronomers began to find

Heinz Pagels.

indications of exotic forms of matter in the universe, and it began to look as if it could be closed: this was required in Tryon's model, though not in Brout's.

Then came inflation theory. It *required* that the density of the universe be virtually equal to the critical density. Tryon's idea soon resurfaced and others began to take an interest in it. In 1981 Heinz Pagels and David Atkatz of Rockefeller University looked at an extension of the idea. They envisioned the universe beginning as a tunneling from the false vacuum. We talked about this vacuum earlier in relation to inflation. Their initial state was a tiny empty pellet with more than four dimensions. An even tinier subspace of this higher-dimensional space then tunneled through to give the fireball. But as in the other theories, there was a problem. They did not explain how the space got extra dimensions, nor why only four remained after the tunneling.

So far all theories had postulated the existence of some sort of "space" before the moment of creation. This, according to Alex Vilenkin of Tufts University, was unsatisfactory. If the universe was to be truly created, then even space itself should be created. In 1983 he proposed that the universe began from nothing. He was convinced that, even though it is difficult to visualize, there was absolutely nothing here before the big bang.

I asked Tryon what he thought about the idea of creation from nothing—in particular, what it meant. He laughed, then after considerable hesitation said, "That has to be one of the deepest questions that has ever been considered by man. It may well be that we will never have a confident answer. I suppose people suggested it because we have no basis for supposing that before our universe existed there was a vacuum of the kind we are familiar with. But I don't think anyone has a clear idea what they mean when they say that before the big bang there was 'nothing.'"

HAWKING'S UNIVERSE

Not only is the concept "nothing" a problem to most cosmologists, but equally difficult to comprehend and explain is the sudden creation of a singularity out of this "nothing." A singularity is, after all, a point where all physics breaks down, and we therefore have no way of mathematically describing it. So how can we describe its creation from nothing?

Stephen Hawking of Cambridge University has spent most of his adult life studying singularities. After establishing several important theorems governing singularities in black holes, he attacked the greatest singularity problem of all: the one associated with the creation of the universe. And although he soon showed that there had to be a singularity associated with the big bang, he is now having second thoughts about it.

Stephen Hawking.

Born in 1942 at Oxford, Hawking's early life was spent at Highgate, North London, where he attended St. Albans school. His interest in science developed early. By the time he was in high school he was already asking himself where the universe came from, and how it was created. He wanted to go into physics and mathematics, but his father, a doctor who specialized in tropical diseases, did not encourage him. He felt that there was little future, and too few jobs, in such an abstruse field. He wanted Stephen to go into biology and follow in his footsteps. But Stephen was independent and had made up his mind. When he went to Oxford in 1959 he applied to the physics department, taking entrance exams in both physics and mathematics.

But, like Einstein, Hawking was far from an ideal student; he frequently skipped classes and rarely took notes when he did attend. Nevertheless, he did exceptionally well on all exams. He seemed, in fact, capable of solving almost any problem that was put before him.

While at Oxford his interest in the universe intensified, and he finally decided that he wanted to become a cosmologist. He therefore applied, upon graduation, to Cambridge University. He hoped to work under the steady state cosmologist Fred Hoyle, but was assigned to Dennis Sciama. And it is to Sciama's credit that he soon realized that he had an outstanding talent on his hands. "He always said 'but' to any statement I made," said Sciama. "He had such a cogent feeling of what we were discussing. This might happen after a couple of years with a bright student—but not after one month."

Hawking had barely started working on his Ph.D. thesis, however, when he noticed he was beginning to stumble as he walked, then his speech became slurred. Something was wrong.

Diagnosis showed that he had a wasting neurological disease—ALS, sometimes called Lou Gehrig's disease. Hawking was severely shaken when he heard, and soon became depressed. What was the use of struggling to finish his degree if he was just going to die anyway, perhaps even before he got it finished. And for several months he did virtually nothing. He admitted later, though, that at this stage it was not just the disease that was depressing him. As strange as it might seem, he was also having problems with his thesis; his mathematical background was still weak, and he was making little progress.

Most people who contract ALS deteriorate rapidly and usually live only a few years. And at first Hawking seemed to fit the profile. He grew increasingly weak, until he could no longer walk by himself. And his speech became almost incomprehensible.

Then something happened that brought him quite abruptly out of his depression. He met an undergraduate student, Jane Wilde, at a party one night. Outgoing and soft-voiced, Wilde was attracted to Hawking despite his obvious handicap, and they were soon engaged. Hawking quickly realized that if he was to be married he would have to complete his thesis and get a job. He was soon hard at work.

They were married in 1967 (and now have two sons and a daughter), and at about the same time Hawking noticed that his disease was beginning to subside. Its rate of progression had slowed considerably. And, although he was now confined to a wheelchair, his mind was still as sharp as ever.

He has apparently taken his misfortune in stride. When asked about it recently he replied, "In one respect it's a blessing. It leaves me a considerable amount of time to just think."

And think he did. In fact, within a few years he had made several important contributions to physics. And he managed this despite the fact that he cannot hold a pen to write an equation. He cannot even turn the page of a book by himself. Any calculations he has to do have to be done in his head. But he almost always works with a colleague and most of the mathematical details are left to him. Hawking says he does not like manipulating long mathematical expressions in his head, but prefers to work with a geometrical picture of the problem.

One of Hawking's first major contributions was with Roger Penrose of Oxford. Together they proved a theorem showing that there had to be a singularity associated with the big bang. This led to an investigation of the singularity and event horizon of black holes. Jacob Bekenstein, a Princeton graduate student, had made an interesting discovery about black holes, showing that they may have a surface temperature greater than 0 K. This seemed unnatural because black holes absorb everything that comes near them, and therefore are perfect absorbers. They should therefore have a temperature of absolute zero.

Hawking found, however, that there were strong stretching forces (called tidal forces) near the event horizon of small black holes, and as a result particle pairs would be generated. In some cases one of the pair would fall into the black hole and the other would escape. The net result would be a black hole that appeared to emit particles and radiation. In effect it would be hot. The effect is negligible in large black holes (a few miles across) but can be quite significant in small ones. In fact, as a black hole emits radiation it losses mass and grows smaller. This, in turn, causes it to emit even more radiation, until finally it literally explodes.

When Hawking first discovered this strange result he could not believe it. It seemed impossible: black holes could not emit particles and radiation. They could not be hot. But his equations told him they could. He was reluctant at first to publish his results, but was encouraged by his colleagues.

So, in 1974, Hawking announced his results. And they soon caused a sensation. But, as might be expected, not everyone was convinced at first; a number of scientists even referred to them as "rubbish." When the dust had finally settled, though, everyone realized that Hawking was right. Not only did black holes radiate, but the formulae he had discovered were a first link between quantum mechanics and general relativity. For years scientists had tried unsuccessfully to bring these two theories together. Hawking's ideas provided a first and important link between them.

Hawking then turned back to the universe. A black hole is, after all, an excellent model of the universe. Like a black hole it has an event horizon, and at the beginning there may have been a singularity. Hawking struggled for several years trying to understand this "universal singularity." But the harder he worked, the more it seemed that the singularity would have to go. There was no way we were ever going to use mathematics to describe it. He finally came to the realization that he had to find a way around it—in short, devise a cosmology that had no singularity.

The key, it seemed, was Heisenberg's Principle of Uncertainty. Because of this principle there is a "fuzzyness" associated with the time $t = 0$. In essence, time is smeared out, and therefore there is no unique moment of creation. In the early 1980s Hawking joined forces with Jim Hartle of the University of California at Santa Barbara to see if he could come up with a satisfactory theory that incorporated this idea. And what they came up with was a "wave function of the universe."

It is well known that in quantum theory a "system" such as an atom is described by a wave function. Hawking and Hartle applied the same idea to the entire universe. This might seem strange, but if you remember that the very early universe was only about the size of an atom, it's not so odd. With quantum mechanics we have the tool, assuming we know the wave function, to see how a system evolves, or changes in time. Hawking and Hartle hoped to do this using the wave function that represented the universe.

I asked Hartle what he thought the probability of their theory succeeding was. "While the theory is compelling in many ways, it has to be tested in a lot of details," he said. "We'll just have to wait for the outcome of the tests. The theory has been remarkably successful so far, but there's a lot more to be done, both in exploring it and coupling it with fundamental particle physics." In regard to the problem of the singularity he said, "The theory gets rid of the singularity in the sense that it deals with a geometry that is nonsingular. Nevertheless, it does predict that the universe has a finite probability to go to zero radius."

So, although the theory is interesting, much more work remains to be done before we will know if it is viable. But it does look promising.

HIGHER-DIMENSIONAL COSMOLOGIES

Another cosmology that may eventually be important in relation to the early universe is one with more than four dimensions. Such theories are usually referred to as Kaluza–Klein theories, named for the two men who developed them. Kaluza was the first to use more than four dimensions; his theory, an attempt to unify electromagnetism with general relativity, contained five dimensions. Kaluza was a Privatdozent in Königsberg—a low-ranking academic who received only the fees paid to him by his students. And in most cases that was very little. Kaluza was, no doubt, very close to starvation about the time he came up with his idea. In 1918 he taught a class in tensor analysis that had only one student in it.

Einstein eventually became interested in Kaluza's work and

helped him get a professorship at Kiel. His paper on higher-dimensional theory was published in 1921, and, as it turned out, was the only contribution he was to make in the field. Klein, a well-known mathematician, read the paper and in 1926 published two papers extending Kaluza's ideas. In the first of the two he merely patched up Kaluza's mathematics, but in the second he made an important addition. He suggested that Kaluza's fifth dimension was a physically real dimension. Kaluza had not done this; he felt that he was merely using five-dimensional arrays to unify electromagnetism with general relativity. Of course, if Klein was convinced the dimension was real, he had to explain where it was in our universe. And he did. Using quantum theory he showed that one of the dimensions would be curled up with a circumference of only 10^{-30} centimeter—so small it would be completely undetectable.

Since this early theory, several similar theories have been put forward. Scientists, for the most part, though, are no longer attempting to unify electromagnetism and general relativity, but rather are hoping that these theories will be useful in explaining creation. The new theories are different from the early ones in that they contain more than five dimensions—in most cases ten or eleven. But as in Klein's case they have to assume that only four appear in the real world, with the others somehow remaining small. These theories are of considerable interest but so far none of them have been highly successful.

REFLECTION

Let us take a few moments to reflect on the material of this chapter. It is certainly easy to agree with Tryon's statement that the origin of the universe is one of the deepest problems ever contemplated by man. How, indeed, was the universe created? Is it even possible for us to find out? We do, of course, have an alternative. We could say that there was no creation, and that the universe has always been here. But this is even more difficult to accept than creation. What, after all, does "always" mean? The idea of an infinity is difficult to swallow in relation to any problem. What, for example, do we mean when we say that the universe extends to infinity? Certainly, it's difficult to know.

If we accept the big bang theory, and most cosmologists now do, then a "creation" of come sort is forced upon us. If the universe was *truly* created, though, there had to be nothing here before the creation. And again this is something that is difficult for most people to visualize, difficult because we usually associate "nothing" with empty space. But as we have seen this is not true; space is bubbling over with particle–antiparticle pairs. "Nothing" can therefore only mean no space and no matter. In short, we have to say that at time $t = 0$ the universe was created, and before that time it did not exist.

Of course, this means that "time" also did not exist. Is this possible? Again. it is an idea that is difficult to comprehend, but no worse, I suppose, than the lack of space and matter. But, as Jim Hartle recently said to me, "Despite the difficulties, the problem of the initial conditions of the universe is one that is not going to go away, and will no doubt become, as our observations in cosmology become more accurate, an increasingly important problem."

It will be interesting to look back on our ideas 100, or even 50 years from now and see how drastically they have changed. We may be in for quite a surprise.

CHAPTER 12

The Big Breakup

We have traced the development of the universe from its origin to the creation of the first atoms. Even while these atoms were being formed, though, the universe was still generally uniform—an expanding gas of particles, nuclei, atoms, and radiation. Eventually, of course, the gas had to break up. We know this because when we look out into the universe today with telescopes we see that it is no longer uniform. We see the stars of our own system, the Milky Way, and beyond it we see billions of other galaxies. The universe is dotted with galaxies. If the expanding gas of the early universe had not broken up we certainly would not see galaxies.

This brings us to the question: How and why did this cloud break up? Or, more generally, how did the galaxies form? (This, incidentally, is still one of the foremost problems of cosmology.) The best answer, according to astronomers, is that they began as small fluctuations in the material of the early universe. They were regions where the density of particles was slightly higher than average. Once these fluctuations developed they would, of course, pull other matter toward them because of their excess gravity. And they would grow eventually into galaxies. This sounds like an excellent solution to the problem. And it is—except for the fact that it leaves us with another problem: Where did the fluctuations come from? How did they arise?

To answer these questions we have to look back to the early universe—to the era before the Planck time. As we saw earlier, this era is generally assumed to have been a froth of disconnected space, made up mostly of tiny black holes popping in and out of existence. But it had an important property: it was inhomogeneous. Was this inhomogeneity responsible for the fluctuations? Many astronomers believe it was. On the other hand, there are other explanations. The fluctuations could have been caused by reactions in the heavy particles of the early universe. As they decayed inhomogeneities could have arisen.

Although there are still many uncertainties in relation to the fluctuations, we are relatively certain that there were fluctuations of some sort in the early universe. The theory of how these fluctuations would have grown is due to the English astrophysicist James Jeans. Born in 1877 at Ormskirk, Lancashire, Jeans was a precocious child, telling time at three and reading at four. His passion as a youth was numbers; he memorized the first twenty logarithms to several decimal places before he went to school. Upon completion of his Ph.D. at Cambridge University he taught at Princeton University for several years. It was during this time that he became interested in a theory of the origin of the solar system that had been put forward many years earlier by the French mathematician Pierre Laplace. Laplace had assumed the early solar system was a rotating disk of gas which broke up into a series of rings. According to the theory these rings eventually condensed to form the planets. Jeans studied the dynamics of such a disk and found that if small fluctuations developed, two forces would determine their subequent fate. Gravity would, of course, tend to pull other material toward the higher density region of the fluctuation, but the pressure within this region would disperse it. Which of the two forces would win out depended on the size of the fluctuation. If it were sufficiently large, gravity would overcome pressure and the fluctuation would grow until it ran out of surrounding material. On the other hand, if it were small, internal pressure would disperse it and it would dissipate. The dividing line between these two cases is now named for Jeans; it is called the Jeans length.

Although Jeans only applied his theory to the formation of

James Jeans.

planets, and later to the dynamics of nebulae, it is also invaluable in explaining the evolution of the fluctuation of the early universe. There was, however, something important affecting these fluctuations that Jeans did not include in his theory. The expansion of the universe severely affected their growth. As matter is pulled toward a region of higher density, the universe expands and disperses it. Indeed, as we will see later, this is not the only thing that impedes it. So, although fluctuations no doubt developed early on, they likely grew very little during the early stages of expansion.

INFLATION

The theory of fluctuations and their subsequent growth into galaxies fell on hard times about 1950. For ten years almost nothing on the subject was published. Then in the mid-1960s interest began to build again, but it was not until the late 1970s that a significant breakthrough occurred. That breakthrough was inflation theory.

As we saw earlier, inflation theory solved several of the problems of the standard big bang theory. And when it was applied to the problem of fluctuations and the formation of galaxies, astronomers were delighted to find that it also partially solved this problem. How? We know that inflation occurred early—about 10^{-36} second after the big bang. And at this time quantum fluctuations would have existed because of the Uncertainty Principle. Inflation merely amplified them. What they looked like at the end of inflation depends, of course, on how long inflation lasted and how extensive it was, but it seems reasonable that they were at least macroscopic in size.

One way of thinking of these fluctuations is as slight "wrinkles" in the geometry of space-time, much in the same way there are wrinkles in a balloon before it is blown up. Inflation merely smoothed out and increased the size of the fluctuations.

One of those involved in using inflation theory to produce galaxies was Paul Steinhardt of the University of Pennsylvania. I asked him about his work. "When we (Turner, Bardeen, and myself) first embarked on computing the spectrum of fluctuations generated by inflation," he said, "we worried that the result would be completely incompatible with the homogeneity observed on large scales. We were excited, though, when we proved the spectrum was just the sort needed." He went on to say that although there are still problems with the theory he is encouraged by the successes.

Andreas Albrecht of Fermilab, who is also working on the problem, says, "I believe the details of the formation of galaxies are something we will understand much better in the next few years. New observations are constantly coming in and there is a lot of exciting theoretical work going on. I expect to see major advances in the area in the next decade."

But what happened to the fluctuations when inflation

ended? Calculations show that they became "locked" into the expansion of the universe, and as a result, they changed very little. The reason for this was radiation. When the temperature was still high (10^{10} to 3000 K) the universe was radiation dominated. Matter and radiation were in equilibrium with one another. As fast as a particle was generated it was absorbed, and as a result the motion of the electrons and protons was inhibited. They were being bombarded on all sides by photons and could not collapse. This is referred to as "radiation drag."

But the universe was cooling rapidly because of expansion. Finally, at a temperature of about 3000 K, protons began to capture electrons, and the first hydrogen atoms formed. As a result the radiation decoupled from the matter and expanded freely into space. The fluctuations were now unimpeded and could grow. At first they continued to expand with the universe, but their own gravity impeded them and they soon began to lag farther and farther behind. Finally they decoupled from the expansion and began to collapse in on themselves.

At first they were just huge gas clouds. But gradually turbulence developed within them and they began to fragment. Then the fragments broke up, until finally the cloud had become lumpy. The lumps then became opaque, trapping most of their radiation. At this stage we refer to the cloud as a protogalaxy. Stars were not yet present, but they would soon form. As the fragments collapsed in on themselves, the radiation pressure increased until it exerted a considerable outward pressure. The collapse continued until the two opposing forces—the inward force of gravity and the outward pressure—finally equalized. The inward fall then stopped and nuclear reactions were triggered in the core of the object. The fragments had become stars.

Galaxies are, as we saw earlier, a relatively recent discovery. Not until the 1920s were we sure they actually existed. Astronomers had seen tiny diffuse regions of light in their telescopes for years, but there was considerable controversy. Kant had suggested that they were systems of stars like our Milky

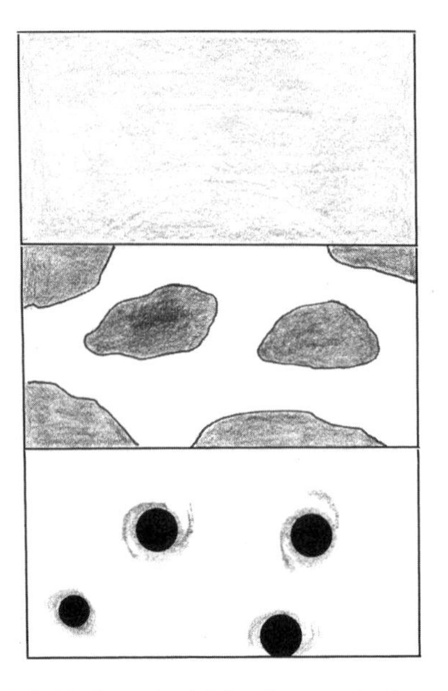

The breakup of the big bang cloud. Time increases in downward direction.

Way. But most astronomers did not agree with him. Some thought they were just huge patches of gas, and a few even thought they might be solar systems in formation. In the mid-1800s Lord Rosse of Ireland, using a 72-inch reflector, noticed that many of them had a spiral structure; arms could be seen emanating from their core. Then in 1914 Vesto Slipher of Lowell Observatory showed that they were rotating like giant pinwheels in space. But still, few were convinced.

Proof finally came from Edwin Hubble. Using the 100-inch reflector at Mt. Wilson he took long exposures of some of the nearby nebulae (as they were known then). The outer regions were finally resolved—individual stars were seen—some of them Cepheids. Using these Cepheids he proved that they were systems beyond ours—galaxies.

A gaseous cloud. The birthplace of stars (Orion nebula). (Courtesy National Optical Astronomy Observatories.)

THE FIRST GALAXIES

What were the first galaxies like? As strange as it might seem, we should still be able to see them. They will no doubt have changed significantly since they were born, but nevertheless we may be able to see them as they were then. I talked about the reason for this earlier. As we look out into space we are actually looking back in time. If you went outside tonight, for example, and looked at the Andromeda galaxy you would see it as it was two million years ago. This means that as we look farther and farther into the universe we see galaxies that are younger and younger. Interestingly, though, we eventually run out of ordinary galaxies like our Milky Way. Beyond the most distant ones we find mostly active galaxies called radio galaxies (with a few exceptions—see below). The cores of these galaxies appear to be extremely active, and may be undergoing an explosion. Finally, in the outermost regions of space we find only extremely active objects called quasars. Astronomers have determined that they are much smaller than galaxies, perhaps no larger than our solar system, yet strangely they give off more energy than an entire galaxy—even more than a radio galaxy.

We are still uncertain what quasars are, but, at first glance, it seems that they are somehow related to galaxies. It seems, in fact, that they might be the first stages of galaxies. After all, we are looking back in time. A logical question is therefore: Do quasars evolve into radio galaxies, which in turn evolve into ordinary galaxies? Most astronomers do not accept this. They believe that there are ordinary galaxies in this region, but they are just too dim for us to see at the present time. There is, however, some indication that quasars may be associated with galaxies. In recent years a number have been found that appear to have a fuzzy region around them—perhaps the first indication of arms.

But if quasars are not the early forms of galaxies, what do these early galaxies look like? There have, in fact, been several searches for them (we now refer to them as primeval galaxies).

One of the first was made in 1967 by Bruce Partridge and Jim Peebles. They did not find any, but their efforts spurred several other groups. In 1985 Hyron Spinrad of the University of California and several colleagues announced they had photographed a group of galaxies at a distance of 10 billion light-years. Spinrad said at the time that within a few years he believed we would be able to see such galaxies at a distance of 17 billion light-years—almost back to the big bang itself. And his prediction may have come true. In 1987 a University of Arizona team of astronomers announced that they had detected two galaxies at a distance of 17 billion light-years.

THE DEVELOPMENT OF GALAXIES

Once galaxies formed it is natural to ask: How did they develop into the types we see today, namely, spirals and ellipticals? (Actually, there are a small fraction that are irregular, but we will neglect them.) Let's consider spirals. As their name suggests, they have two or three spiral arms. Furthermore, they usually contain about 100 billion stars and are, on the average, about 100,000 light-years across. From a distance they have a disk-shaped appearance with a dense core. Our galaxy, the Milky Way, is of this type.

What caused the spiral arms? Before I answer this I should point out that what we see in looking at a spiral is deceptive. It appears as if most of the stars are in the arms, with few between. This is not actually the case. To a first approximation the density of stars around a spiral is uniform. There are a few more in the arms, but the excess is small. What makes the arms stand out is the large number of large blue stars in them, and the luminous gas surrounding them.

One of the first to take up the study of spiral arms was Bertil Lindblad of Sweden. The son of an army officer, Lindblad obtained his Ph.D. at Uppsala in 1920. After teaching at the University of Stockholm for several years he became the director of

The Whirlpool galaxy. (Courtesy National Optical Astronomy Observatories.)

the Stockholm Observatory. Studies had indicated that the stars around the sun appeared to behave strangely. Some of them appeared to be moving faster than us, others slower. Lindblad wanted to find out why. In 1921 he published a paper showing that our galaxy was rotating around a point in the direction of the constellation Sagittarius. And it was rotating much in the same way the planets of our solar system do. We know, for example, that the inner planets revolve faster than the outer ones. Mercury, for example, being the innermost planet, has the fastest speed in orbit. Both it and Venus travel faster in orbit than does Earth; Mars, on the other hand, travels slower.

We now know that Lindblad was correct for the stars around us. But it turns out that our galaxy is different in several respects from the solar system. In the solar system most of the mass is concentrated at the center in the sun. Not so in the case of our galaxy. The core of our galaxy is dense, but its mass is spread through a large volume. The stars are, in fact, close enough to one another that there is a strong mutual gravitational attraction between them. And as a result the entire core moves as a solid. This means that the stars at the center move no faster than those farther out. Once we are out of the core, however, orbital speeds do decrease as we move outward. In other words, there is what astronomers call differential rotation.

Because the arms of our galaxy appear to trail the core, it seems reasonable that they might have been caused by differential rotation. If, for example, we have a pail of white paint, pour a line of black from the center to the outer edge and begin stirring near the center we soon see a spiral appear. Is this the way the spiral shape of galaxies arises? It is easy to show that it is not. Our sun completes a loop around our galaxy about once every 250,000 years, which means that it has completed about 20 orbits since its birth. Our galaxy is, of course, much older than our sun, and it has therefore completed many more—at least 60. And, after that many orbits, even if it started out quite loosely wound, it would have wound itself up tight. In fact, as this would be true of all galaxies, and as galaxies are all about the

Galaxy in Pisces. (Courtesy Kitt Peak National Observatory, Cerro Tololo Inter-American Observatory.)

same age, all of them would be tightly wound. We know this is not the case.

Indeed, as we look out into the universe we would see galaxies that are much more loosely wound than nearby ones (remember, we are looking back in time). And again we do not. We can conclude from this, therefore, that differential rotation cannot account for spiral form.

What are the alternatives? One of the best came from Lindblad. After spending years studying spirals, he noticed in the late 1950s that the arms did not rotate at the same speed as the stars. It was taking twice as long for the spiral pattern to move around as it was the individual stars. This was only possible if some sort of "wave" was passing through the stars. Thus

Lindblad proposed his *density wave theory*. According to it the arms were due to a wave that passed through the stars.

What would such a wave be like? The type of wave you are likely most familiar with is the one that forms when you throw a rock into, say, a lake. It is referred to as a transverse wave, and is not the type that Lindblad visualized as passing through our galaxy. He thought in terms of a longitudinal wave—the type that passes through the molecules of air when we talk. They bunch up in certain regions, and spread out in others. In the same way, as the wave sweeps around the galaxy, gas and dust pile up in certain regions, and are pulled out of others. In the regions where they pile up, gravity causes them to build up even more. And when the gas becomes concentrated enough, stars begin to form—some of them large blue ones. These stars would make the arms stand out.

In 1964 Chia-Chiao Lin of MIT and Frank Shu of the University of California picked up on Lindblad's idea. They also assumed that the arms were due to a density wave, but they concentrated on gas and dust, rather than stars.

A finishing touch was put on the theory in 1966 by M. Fujimoto of Japan. He discovered that the density wave was traveling at three times the speed the gas could conduct a wave. The wave was therefore supersonic (it was traveling at greater than the speed of sound). This, in turn, meant that there had to be a "sonic boom" along its leading edge which pushed the gas clouds together, speeding up star production.

The theory seemed to be a success. It made several predictions that were consistent with observation. Yet there were problems. Many things remained unexplained. For example: What caused the density wave? Also, in 1971, Alar Toomre of MIT showed that such waves could not last more than a few billion years. Yet galaxies were all about 15 billion years old. Another problem was the number of arms. Some spirals had two, others had three, and a few even had four. The density wave theory could not explain this. And there was the ragged structure of the arms in some cases. Spurs emanated from some arms, and there were bridges between arms in some cases. There seemed to be no explanation of this.

In an effort to overcome some of the problems of the density wave theory, another theory was formulated. It is referred to as the supernova model. Near the end of its life a massive star can explode as a brilliant supernova. These supernovae are so bright that we can see them even though they occur in other galaxies. Occasionally a supernova in a nearly galaxy is so bright it rivals the entire galaxy in brightness.

Now for the theory. We know that when a large gas cloud breaks up, stars form from the fragments, some of them quite massive. Let us assume one of these large stars explodes as a supernova. The shock produced by the explosion will likely compress nearby clouds and speed up star production. And some of the stars produced will soon, in turn, explode and affect their neighbors in the same way. Eventually we will have a chain of bright stars extending outward from the core. Most of the stars are massive so they will be long gone before the arms get wound up tight, so there's no problem in this respect. Such explosions could easily explain spiral arms, spurs, and bridges and perhaps some of the other problems mentioned above.

One final mechanism that also no doubt produces structure in galaxies is collisions. The stars within a galaxy rarely collide, but the motions of galaxies themselves are not nearly as systematic as the motions of the stars within them. Galaxies do occasionally collide. Individual stars do not physically collide but the mutual gravitational attraction between the two galaxies can pull stars out in long arcs. Computer simulations of such collisions show that this likely occurs.

Finally, a few comments on elliptical galaxies. Ellipticals are different from spirals in that they have no arms, and consequently no gas or large blue stars. They are composed mostly of old stars. They also rotate slower than spirals. One theory for their appearance relates to this rotation. Because ellipticals rotate relatively slowly, stars formed in them early, and most of the large ones soon exploded as supernovae. At the present

Spiral galaxy in Ursa Major. (Courtesy National Optical Astronomy Observatories.)

The elliptical galaxy M87. (Courtesy National Optical Astronomy Observatories.)

time, therefore, only small stars exist in them. No new stars are being formed because there is no gas. This is, of course, consistent with observation.

Star formation in spirals, on the other hand, was inhibited in the outer regions because of their fast rotation. The gas remained diffuse and stars formed slowly over a much longer period of time. And, indeed, they are still forming.

SUMMING UP

So far our story of the evolution of the universe has brought us to the point where the expanding gas cloud of the early universe begins to break up. We are still uncertain why it breaks up, but as we saw, inflation theory tells us that the fluctuations that produced galaxies were a result of quantum fluctuations that existed in the Planck era. These fluctuations were locked into the expansion of the universe for thousands of years because of radiation pressure. But when the radiation decoupled from the matter they broke away and began contracting.

As we look at the galaxies today we see that they are of two major types: spiral and elliptical. The spirals, we believe, are a result of a density wave that sweeps around in them. Ellipticals, on the other hand, are so shaped because their relatively slow rotation allowed stars to develop early.

In short, we now believe we know generally how galaxies formed and evolved, but many problems remain. In the next chapter we will look at one of them.

CHAPTER 13

A "Lumpy" Universe

By the mid-1970s astronomers believed that they had a fairly good idea how galaxies formed and evolved. We will see, however, that there were still surprises to come. Hubble had established in the 1930s that galaxies were fairly uniformly distributed across the sky. And since then it had generally been accepted that on the galactic scale the universe was homogeneous.

Closer up, though, it soon became clear that this was not the case. Fritz Zwicky showed in the late 1930s that galaxies have a tendency to cluster. And in the early 1940s he discovered the huge cluster of galaxies in Coma Berenices. What about higher-order clusters—clusters of clusters? Did they exist? There were hints that even this was possible. The Swedish astronomer Charles Charlier, as early as 1908, talked about a hierarchy of the universe: a universe consisting of clusters, clusters of clusters, and so on. Of course, at that time we still had not discovered galaxies.

The idea that there might be clusters of clusters was occasionally discussed, but few took it seriously. Then came Gerard de Vaucouleurs, a French astronomer with imagination and determination. After graduating from the Sorbonne in France in 1957 he and his wife (also an astronomer) went to Mount Stromlo Observatory in Australia to work on a catalogue of bright galaxies. Using 30- and 74-inch reflectors they photographed and studied the southern galaxies. It had already been established that our galaxy, the Milky Way, was in a group of about 20 galaxies called the Local Group. The Andromeda galaxy, the large galaxy in Triangulum, and the Magellanic clouds were all part of this group, along with numerous dwarf galaxies. The Milky Way was second largest, Andromeda being the largest.

Beyond the Local Group were other groups—one in Hercules, one in Virgo, and another in Coma Berenices. They were much larger than ours; the Virgo cluster, for example, consisted of several hundred galaxies. As de Vaucouleurs studied these groups he came to the realization that we were part of a cluster of galaxies—what he referred to as a supercluster. At the center of this supercluster was the Virgo cluster. Overall it had a diameter of about 100 million light-years, and was generally shaped like a pancake, with a thickness of about 6 million light-years. We were on the outskirts of the supercluster, located about 50 million light years from Virgo.

de Vaucouleurs published paper after paper giving details of his model. He even named it the "Local Supercluster" after the name we give to our local group of galaxies. But almost no one paid any attention to him. Most believed that he was seeing an illusion. But the rejection did not deter him. He continued to push his model at meeting after meeting, in paper after paper. Finally, after about 25 years, astronomers began to realize he was right.

But even before de Vaucouleurs had discovered the Local Supercluster, two extensive surveys of the sky had already begun. Both were initiated shortly after World War II. The first was made with a 20-inch telescope at Lick Observatory. C. Donald Shane, director of the observatory, along with Carl Wirtanen, an assistant, photographed the entire sky visible from Lick. Twenty years later they were going to rephotograph it and check to see how the stars had moved. But the plates were dotted with hundreds of background galaxies, and Shane hated to see such a goldmine ignored; he therefore decided to study them himself. Working with statisticians Jerzy Neyman and

Virgo cluster. (Courtesy National Optical Astronomy Observatories.)

Elizabeth Scott of the University of California he analyzed their distribution and found that they tended to cluster. Unfortunately, he could not prove the clustering beyond a doubt. The major problem was that in the photographs the clusters were seen in only two dimensions. Galaxies that were in the same direction but at different distances were superimposed on one another.

The second survey of the sky, which was funded by the National Geographic Society, was made at Palomar using the 48-inch Schmidt telescope. One of those working on the survey was George Abell of UCLA. Abell soon became intrigued by the large number of clusters of galaxies in the photographs and decided to catalogue them as part of his doctoral thesis. His catalogue, which gave the positions of 2712 clusters of galaxies, was published in 1959. In it Abell pointed out that many of the clusters were grouped into superclusters.

But at about the same time Fritz Zwicky of Mt. Wilson Observatory made up a similar catalogue based on the same survey. And strangely, Zwicky pointed out that (according to his survey) clusters did not tend to form higher-order clusters.

Who was correct? The impasse was eventually resolved when it was realized that Zwicky had catalogued generally larger and more diffuse clusters than Abell had. Furthermore he had included several regions where galaxies were highly concentrated.

ONE MILLION GALAXIES

About five years after the Shane–Wirtanen catalogue was published it came to the attention of Jim Peebles of Princeton University. Peebles realized that fast computers could be used to extend the work. He decided to do a detailed mathematical analysis of the clusters to find out once and for all whether the clumping was valid.

Born in 1935 in Winnipeg, Manitoba, Peebles got his B.S. at the University of Manitoba. In 1958 he went to Princeton for his Ph.D., remaining there as a faculty member after he received his degree. At the time he became interested in clustering, he had already made important contributions to cosmology: he was associated with the discovery of the cosmic background radiation and had written a landmark paper on nucleosynthesis.

Peebles was skeptical of superclusters at first. But he soon changed his mind. Using a statistical method called the correlation function method that furnishes the likelihood that a given galaxy has a near neighbor, Peebles and several colleagues proved that galaxies do indeed cluster. This meant that the initial fluctuations in the universe could not have been random. As part of the study Peebles, along with Michael Seldner, Bernard Siebers, and Ed Groth made up a plot of the Shane–Wirtanen

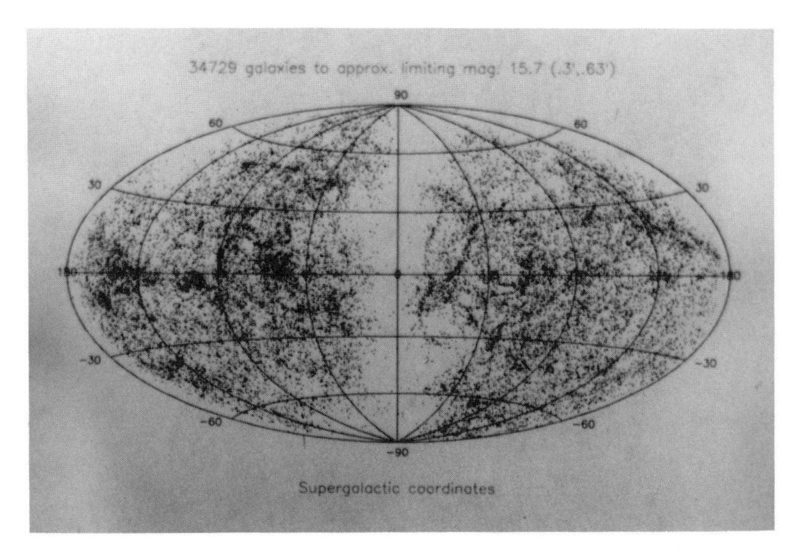

Plot showing the distribution of galaxies. (Courtesy National Optical Astronomy Observatories.)

galaxies. They plotted them according to a shade of gray that depended on the number of galaxies at a given point: dark gray if there were several galaxies at the point, light gray if there were few. The result was a spectacular two-dimensional plot of most of the known galaxies, a plot that soon became a popular poster. Because it contained about a million galaxies it was given the name, "One Million Galaxies." Its most striking feature was its structure: filaments, or series of strings joined at knots. And between the filaments were regions that contained relatively few galaxies. The chains of galaxies were thousands of light-years long.

The filamentary structure and low-density regions were a surprise, but were they real? It was difficult to tell. The plots were two-dimensional, and as a result we were seeing the clusters superimposed on one another. What did they look like in three dimensions? It was not going to be easy to find out. Peebles had plotted a million galaxies. Obtaining the distances to such a large number would be a lifelong task. The distance to individual galaxies and clusters could easily be determined using the Hubble plot. All that was needed was the redshift of the object. But it had taken Humason several decades to get the few hundred he had obtained. Obtaining the redshifts of thousands of galaxies and clusters was almost beyond comprehension.

Further evidence for superclusters and large voids continued to come in. Herbert Rood and Thornton Page of Wesleyan University made a detailed study of the Coma cluster in 1972. A few years later William Tifft of the University of Arizona and Stephen Gregory of Bowling Green University discovered that it had an armlike projection—a bridge of galaxies to

Cluster of galaxies in Hydra. (Courtesy National Optical Astronomy Observatories.)

a nearby cluster. The bridge looked like a long string of beads; it was over 70 million light-years long. Also, there appeared to be a region between us and the Coma cluster that had no galaxies in it—a void. A year later the same researchers located another supercluster about 500 light-years away in Hercules.

Then in 1981 Tifft, Gregory, and Thompson identified another huge supercluster in the direction of the constellation Perseus, about 200 million light-years from Earth. It was so large it extended across Perseus into Pisces, spanning a total of 40 degrees in the sky. About the same time Robert Kirshner, who was then at the University of Michigan, Gus Oemler of Yale, and Paul Shecter of Kitt Peak Observatory discovered a huge void in the direction of Bootes. It was 250 million light-years in diameter, and was, by far, the largest void seen until then.

Astronomers were beginning to get impatient for a more complete survey—a large-scale redshift survey. Peebles had sent out the call and, finally, in the late 1970s it was heeded.

THE FIRST REDSHIFT SURVEYS

By the late 1970s instrumentation had significantly improved and it began to look as if a large-scale redshift survey might be possible. Hubble and Humason's work had been done using slow photographic plates. But now astronomers had an amazing array of new detectors and other devices. Image intensifiers were being used to brighten images. Photon counting devices so sophisticated they can count single photons were also now in use. And a microchip referred to as a CCD (charge-coupled device) had been invented. This device converts photon intensity at each region of a plot into a charge and stores it. Later, the charge at each point is read off using a computer.

"The information and data capability has expanded tremendously with the advent of modern detectors," said John Huchra of the Harvard–Smithsonian Center for Astrophysics. "I can now, with a sixty-inch telescope, get a redshift in half an hour that took six nights on the hundred-inch when Humason was working with it." Huchra went on to tell me about the dramatic increase in the number of measured redshifts over the years. "In 1956," he said, "there were only about 500. By 1970 this had increased to 1500. Between 1975 and 1980 about 8000 more were added until now we have somewhat over 24,000." Huchra and Marc Davis, now of the University of California at Berkeley, were the first astronomers to be involved in a large-scale redshift survey.

Huchra was born in Jersey City, New Jersey, and went to MIT after graduating from high school. "I think my real start in physics and cosmology came from reading popular books in the 1950s," he said. "I read Gamow, Hoyle, and many others. I was fascinated by them and went to MIT to study to become a mathematician or physicist. I ended up getting a degree in physics, but I did some experimental work in astronomy. I also did some theoretical work. In fact, I did a theoretical thesis on stellar pulsations in RR Lyra stars. When I graduated I went to Caltech, where I assumed I was going to continue doing theoretical work. But at Caltech I began using some of the large nearby telescopes." A smile came over his face. "And I soon fell in love with them," he said. Then, nodding, he added, "That's a dangerous thing. Once that happened it was all over. My fate was sealed. I knew I was going to be an observational astronomer."

Huchra did his Ph.D. thesis under Wal Sargent but said he spent a lot of time talking to Leonard Searle and Allan Sandage of Palomar. "I learned a lot from both of them," he said. His first project in cosmology was on stellar populations in galaxies. But he had heard of Peebles's work and knew that there was a need for a redshift survey.

"About 1975 a group of us at Caltech got together to try to start a redshift survey," said Huchra. "Using the sixty-inch at Palomar, some telescopes at Kitt Peak, and the radio telescope at Arecibo we started taking data. But it went extremely slowly. We were still using the old detectors. Then I went to Harvard in 1976 after finishing up my Ph.D., where I met Marc Davis. He

John Huchra.

was avidly interested in doing a redshift survey." Huchra paused for a moment to collect his thoughts. "Anyway, for one reason or another the collaboration at Caltech broke up and Marc and I started doing work in the later part of the 1970s."

Marc Davis did his graduate work at Princeton University. He worked with Peebles and Dave Wilkinson, both of whom had been involved in the discovery of the background radiation. His Ph.D. project was a search for primeval galaxies. It was a lot of hard work and disappointment, and in the end he could not say with any certainty that he had detected a primeval galaxy. When he graduated he went to Harvard as an assistant professor. Huchra was there as a postdoctoral. He was still feeling a little let down as a result of the primeval galaxy search and wanted to do something that would bring more definite results. A redshift survey seemed to fit the bill.

Within a few months Davis and Huchra had initiated their survey. Using a sixty-inch telescope at Mount Hopkins in Arizona they began taking data. But the instrumentation on the telescope was old and they soon realized that something had to be done if they were ever going to finish.

"I had a friend, Steve Shectman at Mount Wilson, who had made some advances in the detector field," said Huchra. "I convinced Marc we should try to copy his equipment . . . his detector and so on . . . and put it on our telescope. Marc agreed and went off to get the information we needed to build the new equipment. I remained at Cambridge reducing the data we already had obtained."

The changes—when they came—made a tremendous difference. But they did not come quickly. They had to completely rebuild the spectrograph, build a completely new detector, then install a computer to run the system. In 1978, however, the new system was finally ready. "Once we got the new equipment attached to the back end of the telescope we began taking data in reckless abandon," said Huchra. "It took us three to three and a half years to finish the first phase of the survey, which was to do 2400 galaxies in the northern hemisphere." The result was the first three-dimensional map of the intergalactic universe. "We were quite excited when we first looked at it," said Huchra. The filamentary structure that was visible in the earlier two-dimensional plots was still there. Strings of galaxies surrounded voids in the sky. Overall it looked like a giant sponge.

PANCAKES AND PICKLES

The idea that the universe was populated by huge chains of clusters was now well established. And almost all astronomers accepted the voids between them as genuine. But how and why did such a strange structure occur? The answer, it seemed, lay in the early universe.

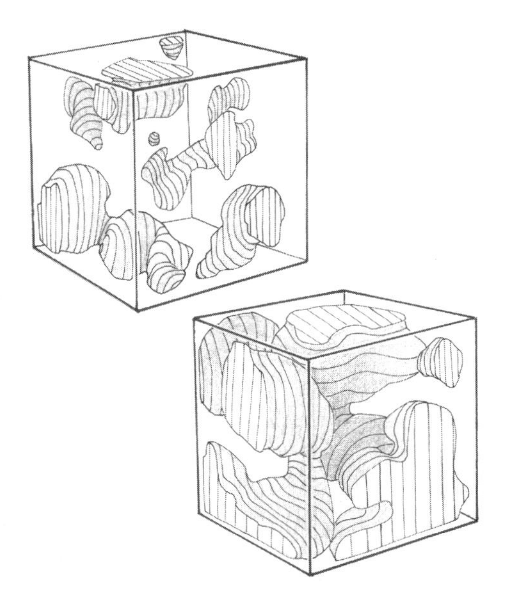

Computer simulation of growth of voids in the universe.

Shortly after Peebles began his work on superclusters he introduced a theory to explain the strange structure he was getting. He assumed that galaxies formed out of the huge gas cloud that emerged in the big bang explosion, much in the way we described in the last chapter. This gas cloud broke up as a result of fluctuations, and each of the individual clouds collapsed to give a galaxy. The universe was still relatively homogeneous at this stage. Some of the galaxies then began to attract other galaxies and a clumping began. First clusters of galaxies formed, then clusters of clusters—superclusters.

Sounds great. But is it correct? If it is, it means that galaxies are relatively old. And, of course, we believe that they are. But clusters formed after galaxies so they would be relatively young, and superclusters even younger. This is the problem. It takes a long time for gravity to pull galaxies into clusters. And, from all indications, not enough time has passed to get the distribution of superclusters we see.

What about the voids? How does Peebles's theory explain them? It does, indeed, explain them in a reasonable way. It assumes that they were left when the galaxies that were originally in them were pulled into clusters. In short, they are what was left behind when the chains formed.

Because it starts with galaxies, which later form clusters, then finally superclusters, Peebles's theory is referred to as the "bottom-up" theory. In sharp contrast is another theory that was formulated in the early 1970s by Y. B. Zel'dovich and several colleagues of the Institute of Applied Mathematics in Moscow. Their theory is referred to as the "top-down" theory. When it was first put forward the filamentary structure of the universe was still in question, and as a result few astronomers paid any attention to it.

Zel'dovich assumed that the initial gas clouds were the size of superclusters. Calculations showed him that if these clouds started to collapse along one axis faster than the others (and they likely would), they would soon become huge "pancakes." These pancakes would be tens of millions of light years in length. Galaxies presumably formed when turbulence within them broke them up. According to Zel'dovich most of the galaxies formed in the dense regions where the pancakes interacted with one another. This is, in fact, what caused the filamentary structure and voids of the universe.

Zel'dovich's original theory was interesting but flawed. The early universe, we know, was very homogeneous. The cosmic background radiation tells us this. How could the original cloud have gone from uniformity to huge pancakes, then to clusters and superclusters so fast? Again, as in Peebles's theory, calculations showed that there was not enough time since the big bang.

The theory seemed doomed. But then about 1980 a number of physicists began speculating that neutrinos might have mass (prior to this time they were thought to be massless). Zel'dovich seized on this to make adjustments to his theory. He and his coworkers showed that the neutrinos would decouple from the big bang plasma well before other particles. And as they cooled they would have been pulled into clumps as a result of their mass. At this stage ordinary matter would still be hot and uniform. It would not begin to clump until the photons decoupled in another 100,000 years. By this time, though, the neutrino clumps would be well established and the matter would therefore be attracted to them. It was these clumps that eventually gave us clusters of galaxies. According to Zeldovich's calculations the size of the clusters comes out about right.

But where are the neutrinos today? According to Zel'dovich they form a halo around the galaxies. Massive neutrinos, in fact, solve another problem, referred to as the "missing mass" problem. It is well known that clusters of galaxies do not have enough mass to gravitationally hold themselves together. Zwicky discovered this in the early 1930s when he was studying the Coma cluster. According to his calculations the Coma cluster should be flying apart. But it was not. The only explanation he found was that there was "dark matter" in the cluster—material we could not see or photograph. It was soon discovered that most other galaxies had the same problem.

But even with this, Zel'dovich's theory was still not out of the woods. According to Peebles, "His theory is dead. Our galaxy is much older than the Local Supercluster. The theory doesn't explain this."

LATER REDSHIFT SURVEYS

When Huchra and Davis finished their survey in the late 1970s Davis left Harvard and went off to a professorship at Berkeley. Huchra decided to continue with the next phase of the work. "We had the telescope and equipment," he said. "I decided we may as well continue on. Margaret Geller, who had worked with us earlier but spent most of her time in England, came back and we formed a collaboration to continue. We wanted to go deeper and do more specialized things. But more than anything we wanted to do some area of the sky a factor of two in distance deeper. We started with clusters of galaxies, primarily because you have to publish papers on time scales of a year or so, or you don't get funding. After we finished all the main clusters and analyzed them we went on to strip surveys."

He stopped for a moment. "Let me explain," he said. "There are a lot of different types of surveys you can do: random surveys of the entire sky, or you can take a small area and drill it as deep as possible. You can also take a large area and do it completely . . . or easier, you can take several intermediate-sized areas and do them, then add them together. And there are strip surveys; they are along strips in the sky. We decided it was best to do this type. While we were working on our first strip survey we just let the data accumulate. We didn't plot it up until we were finished." Much of the plotting was done by Valerie de Lappert, a graduate student who was working on her Ph.D. thesis under Geller and Huchra.

"When we finally plotted the data I almost fell off my chair," said Huchra. "The results were startling. They were not at all what I had expected. Past observations had indicated that we were going to be seeing filamentary structures—connecting clusters of galaxies. That's not what we saw. We saw things that looked more like the surfaces of bubbles . . . the galaxies seemed to be on these surfaces. And that's quite different from filaments. You can make filaments [theoretically] quite easy, but surfaces of bubbles are much tougher."

I asked Huchra which of the two theories—the top-down or the bottom-up—best fitted his observations. "Neither," he said emphatically. "Right now what looks like the best qualitative bet are the explosive scenarios." The best known of these is the one by Jeremiah Ostriker of Princeton University and Lennox Cowie now of the University of Hawaii, and S. Ikeuchi of Tokyo. They have suggested that the energy is released in the formation of early supermassive stars. In short, they imagine supermassive

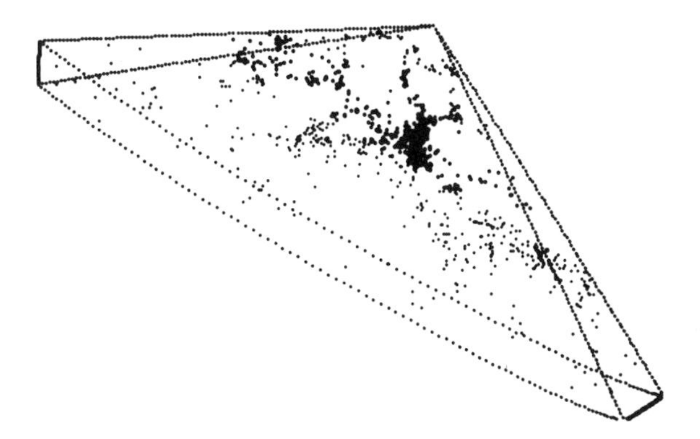

Large-scale plot of clusters. Note bubbles. (Courtesy John Huchra.)

stars that exploded as supernovae, sending out a shock wave into the intergalactic medium. These shock waves scoured out certain regions of space and made others more dense. New galaxies would form in the dense regions and the process would continue.

Huchra described another possibility to me. "It's also possible that the universe has the right type of properties for explosions to be caused by regions of negative energy," he said. "Suppose you have a uniformly expanding medium—such as the plasma in the early universe—and you take a region and empty it out. That region will expand faster than the rest of the universe, the reason being that there is less gravity inside it because it's empty. The bubble will therefore expand. It's possible to get bubblelike structures from a region such as this—one that is underdense in the early universe." He hesitated. "Maybe it would work . . . but on the other hand, maybe it wouldn't," he said, shrugging.

I asked Huchra about how much of the universe his survey had covered. He laughed. "We've now reached to a depth only about three or four percent the distance to the horizon," he said. "We've surveyed a volume that is only a tiny fraction of the overall universe. Borrowing an analogy of Margaret Geller's I like to say that the volume of all surveys done so far including our own, compared to the whole universe is what the area of Rhode Island is to the area of the entire Earth."

What will we see as we probe even deeper? How were the bubbles formed? A possible answer is given in the next chapter.

CHAPTER 14

Cosmic Strings

We have seen that the universe, on the very large scale, consists of filaments separated by huge voids. And looking closer we have discovered that the voids look like bubbles. In fact, they look like the cavities that would be created in an explosion, and as expected, this was the first idea that was put forward. But further examination indicated that there was not enough energy in thermonuclear explosions. It seemed as if something more was needed. In this chapter we will see that there is another possibility.

Earlier we talked about the phase transitions that occurred in the early universe. These transitions are similar to the ones that water undergoes as it cools. You know, for example, that steam condenses to liquid water, and liquid water, in turn, freezes to ice. In the early 1970s Thomas Kibble of Imperial College in London began studying the phase transitions that occurred in the early universe. He soon realized that, just as defects form when water freezes, so too would defects form during these transitions. You are likely familiar with the defects that form when water freezes. If you watch the freezing closely you see that ice first forms in clumps throughout the water. We refer to these clumps as domains. If you look carefully at an individual domain you will see that the ice crystals within it are all aligned in the same direction; in a different domain, however, they will be lined up in a different direction. As these domains merge, defects will therefore form along the merging line.

Kibble realized that the same thing would likely occur in the early universe. Indeed, as a result of inflation theory, we now feel we have a reasonable understanding of the process. We saw that before inflation the vacuum of the universe was very energetic; it was, in fact, what we refer to as a "false vacuum." At this stage the universe was very symmetric; the forces of nature were unified and there was only one type of particle. But then a transition to the true vacuum occurred. The true vacuum emerged in the form of bubbles, and just as the domains of ice came together, so too did the bubbles of true vacuum. And again at the interfaces defects occurred. The defects in this case were trapped regions of false vacuum. And because the false vacuum was extremely energetic, these regions were correspondingly energetic.

In 1976 Kibble described the types of defects that could occur. He concluded that three types were possible. If the surfaces of the bubbles came together they would form domain walls. If the bubbles merged along lines they would form cosmic strings. And finally, if they touched only at points, they would give monopoles. Zel'dovich pointed out immediately, however, that it is unlikely that domain walls exist. He showed that they would be so massive they would severely perturb the microwave background, and we see no evidence of this. Actually, since Zel'dovich's paper, we have an even better argument. Scientists have shown that a single domain wall across the universe would have a greater mass than all the matter now in the universe. We could hardly miss such an object. Because of this, there is now general agreement that domain walls do not exist.

What about monopoles? There has, in fact, been considerable interest in them lately. And there are good arguments for their existence. Particles of electrical charge exist; why not particles of magnetic charge (in other words, monopoles)? Grand unified theories even predict them. But the trouble is that they predict too many. We should be seeing them routinely and we do not. On the basis of this it is best to assume that they do not exist. Or perhaps there are so few of them we'll never observe one. (We saw earlier that new inflation also forbids them.)

The zoo of cosmic defects.

That leaves cosmic strings. Do they exist? A number of scientists are confident that they do. Of the three possibilities discussed above they are, indeed, the best bet at the present time.

PROPERTIES OF STRINGS

Because of the recent interest we now have a pretty good idea what cosmic strings would look and act like. Before I talk about them, though, I should point out that we are still not

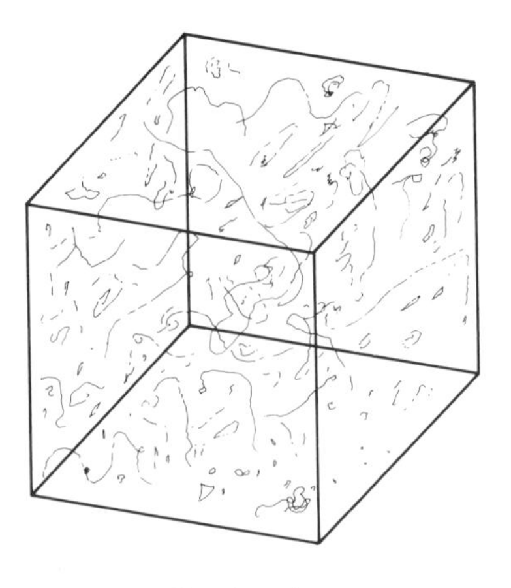

Cosmic strings.

certain they exist. At times I may talk about them as if they are objects of everyday experience, but they are definitely not. Even scientists who are tremendously enthusiastic about them will admit that they are speculative. If they do exist, though, we will see that they may solve several important problems of cosmology.

What would such a string look like? Calculations show that they would be extremely long; in fact they would have no ends. This means that only two types can exist: those that extend from one end of the universe to the other, and those that form closed loops.

How massive would they be? They are, of course, tubes of trapped false vacuum, and because the false vacuum was tremendously energetic, we would expect them to be correspondingly massive. Calculations show that a section an inch long would weigh as much as the Earth. The most massive ones would be those that were formed earliest; they would be the ones with the most energetic vacuum in them. Furthermore, scientists have shown that the earliest ones would be the thinnest. They would have a diameter of only about 10^{-30} centimeter (which is considerably less than the diameter of a proton). It is these thin, massive ones that we are most interested in, for as we will see later, they are the ones that are believed to be associated with clusters of galaxies.

The energy that is tied up in a string also gives a measure of its tension. This tension is, in fact, so great that they would tend to oscillate rapidly, their speed at any point being close to that of light.

VILENKIN

After Kibble published his pioneering paper on cosmic strings there was an interval of about five years in which almost no papers on the subject appeared. Zel'dovich and his colleagues in the Soviet Union continued to work on them, but they published nothing.

The thread was picked up by Alex Vilenkin, who had emigrated to the United States from Russia. Vilenkin published one of the first papers after Kibble, and has since produced a steady stream of papers on the subject. He is now considered to be one of the leaders in the field. Born in Kharkov, USSR, in 1949, Vilenkin got his early schooling in Kharkov, and his undergraduate degree from the University of Kharkov. Although he developed an early interest in cosmology, there were no courses offered at the university in either general relativity or cosmology. Furthermore, no one on the faculty was interested in them. Vilenkin therefore began studying the subjects on his own. Eventually, though, he found someone at a nearby research institute who was interested in them and they began working together. "I visited him about twice a month and he gave me various problems to work on," said Vilenkin.

When Vilenkin graduated he soon learned the realities of Soviet life. He could not get a job in physics—not because none were available, but because he was Jewish. Furthermore, he could not get into a graduate school. "Quite a few professors

Alex Vilenkin.

wanted me to work with them," he said, "but in Russia that has to be cleared with the KGB." And because he was Jewish he could not get it cleared.

For the next few years Vilenkin worked at odd jobs, none of them related to physics. But he was still determined to become a physicist, and as he worked he continued to write papers, many of which got published. He soon realized, though, that if he was to get an advanced degree and work in physics he would have to get out of the Soviet Union. And he knew that would not be easy. Nevertheless, he went ahead and applied for an emigration visa to the United States.

While he was waiting for a reply he met Mark Azbel, a well-known Soviet physicist who had also applied for a visa, but had been turned down. "He was interested in biopolymers, so I got involved in biophysics . . . although I kept thinking about cosmology," said Vilenkin. "It's best to have somebody to talk to about your work . . . and I was fortunate to be able to work for him. I learned a lot from him."

The waiting, fortunately, was not in vain. Vilenkin did eventually manage to get out of the Soviet Union. Upon reaching the United States he went to the State University of New York at Buffalo, and within a year had his Ph.D. And soon he was publishing regularly. One of his first papers, interestingly, listed his affiliation as the Kharkov Zoo. I asked him about this. "I didn't submit this paper until I was in the United States, but it was done while I was working as a night watchman at the zoo, so I put their address on it," he said.

Vilenkin is now at Tufts University. For a while he continued his work on biopolymers. In fact, he did his Ph.D. on the subject. But he soon turned back to his first love, cosmology. "I didn't do my thesis in cosmology because I was sure I wouldn't be able to get a job in the area," he said. But his position at Tufts did, indeed, allow him to get back into cosmology.

Vilenkin's first project in cosmology was a look at the origin of galaxies. He read Kibble's 1976 paper and became interested in cosmic strings. And in his first paper (which appeared in the early 1980s) he showed that strings might indeed be important in relation to the origin of galaxies.

There was, however, a problem related to strings: they appeared to be in conflict with new inflation theory. As I mentioned in an earlier chapter, new inflation gets around some of the problems of old inflation by assuming that few bubbles formed in the early universe. In fact, it assumes we may now live in the interior of what has always been a single bubble. If so, there would be few strings around, as they form where bubbles merge. I asked Vilenkin about this. "I wouldn't say there is a conflict," he said. "There are models in which strings are formed after inflation. Besides, at the present time there is no satisfactory model of inflation. Most cosmologists feel there is a ring of truth to the idea, so something of that sort probably will be included in the final theory. But the specific implementation of inflation is very sensitive to particle physics, and particle physics is still changing rapidly."

He stopped for a moment to collect his thoughts, then continued. "Inflation can be thought of as occurring at a phase transition. And cosmic strings form at phase transitions. But these transitions can be different. They can be independent events . . . so the bubbles responsible for cosmic strings are not the same bubbles that are responsible for inflation."

Vilenkin then went on to describe the oscillations of strings to me. As I mentioned earlier they are under tremendous tension and will oscillate rapidly. What would happen if, during these oscillations, one piece of the string happened to cross another? Calculations show that the strings will break at the point of intersection. So if the original string was in the form of a loop, there would now be two loops. These new loops will, in turn, oscillate, and if they cross, smaller loops will be generated. The loops will therefore get smaller and smaller as time passes. Indeed, this is not the only thing that causes them to shrink. A string is extremely massive, and as it oscillates it gives off gravitational waves. These are waves, similar in many ways to electromagnetic waves, that are produced by oscillating masses. They are predicted by Einstein's theory of relativity, and although scientists are quite confident these waves exist, they have not yet been found.

Of course, when something releases radiation (or equivalently, energy) it must decrease in mass. This means that as the strings oscillate they must shrink. In fact they should get smaller and smaller until they eventually disappear. Calculations have verified that many of the loops that were originally in the universe would have disappeared by now. But some should remain.

GRAVITATIONAL FIELD AROUND A STRING

Because of their high mass, strings are surrounded by a strong gravitational field. And since gravitation is equivalent to curved space, we can say that the space around a string is curved. What would the effect of this be? Certainly, if we were looking at an object behind the string it would appear distorted.

To understand the nature of this distortion let us consider a simple experiment. If you have an ordinary string and draw a circle around it you can write down a number giving the area of the circle. Simple enough . . . the space is flat. But what if we did the same thing around a cosmic string? It turns out that, because of the curvature of space, the area would be larger. The best way to visualize this is to imagine the circle is drawn on a piece of paper. Suppose you cut out the circle, then cut a small wedge out of it (i.e., with sides extending to the center). If you glue the two remaining sides back together you get a cone. This cone gives a simple representation of the curvature of the space around the string.

What, then, would something behind the string look like? Because of the curvature of the space, rays from the object would split, and we would see two images of the object. Such a phenomenon was predicted years ago by Einstein; it is called a gravitational lens. And we have recently found several good candidates for such a lens. We see double and triple quasars, side by side, that appear to be images of the same object. It is not likely, however, that they are caused by cosmic strings because the same phenomenon occurs when any dense object, say a black hole, or even a dense galaxy, lies between us and the quasar.

There is a way, however, that we could tell if the object between us and the background was a cosmic string. Assuming there were objects in the background we would see several double images. In fact, if the cosmic string was in the form of a loop, we would see double images all around the loop. So far we have not found double images of this type, but searches continue.

NETWORKS AND COMPUTERS

What would the cosmic strings of the early universe look like if we could see them? Most people working in the area agree that they would be in the form of a network: loops, filaments,

Andreas Albrecht.

single strings, knots, and so on. One of the important questions today is, in fact: how does this network evolve? Stated another way: if we knew what it looked like early on, how did it change as the universe expanded?

Andreas Albrecht of Fermilab, in conjunction with Neil Turok of Imperial College in London, has been studying this problem for several years. Albrecht was born in Ithaca, New York, in 1957. "Through a good part of high school my dream was to become a professional violinist," he said. "But by the time I was a senior I had developed a keen interest in physics so I went to Cornell and majored in physics." After graduating from Cornell he went to the University of Pennsylvania, where he worked on inflation under Paul Steinhardt. And, we saw in an earlier chapter, he was the coauthor of "new inflation."

From the University of Pennsylvania Albrecht went to the University of Texas at Austin as a postdoc, then to another at Los Alamos. He finally ended up at Fermilab. Albrecht and Turok decided to write a computer program to study the evolution of strings; they wanted to follow the changes in the strings as they aged. They hoped to prove an idea that Kibble had put forward earlier called "self-similarity." It says that as a network evolves it will stay more or less the same, except for size. "Our real hope is to show that no matter how a network starts out it will eventually approach a certain [standard] network," said Albrecht. "That's interesting and important for a number of reasons. The most important is that we can't possibly evolve the computer program from the time the network was created until now. One of the characteristics of self-similarity is that if you have a picture of the network at one time and want to know what it looks like at a later time, you just sort of scale it up—like you blow up a photograph. That makes the problem easier. So it's an important thing to show."

I asked Albrecht about the possibility of detecting cosmic strings. He said, "The best way would be to see if there is a jump in the temperature of the cosmic background radiation in an area where you expect a string. The effect on the photons that fly on opposite sides of the string and eventually reach our eye is to give an apparent shift in temperature. Measurement of such a shift would be important. I feel it's our best bet for detecting them."

SUPERCONDUCTORS

We have talked about the properties of cosmic strings but so far I have said nothing about how they might account for the huge bubbles that have been observed.

The breakthrough that led to our present ideas was made by Ed Witten of Princeton University. Witten was considering how cosmic strings might be observed when he discovered that certain charged particles would have no mass in the high energy vacuum of the string. This meant that particle pairs could easily be generated, and that the two particles would travel in opposite directions along the string. Only a slight energy would be needed to get them moving, and once they began moving they

Jeremiah Ostriker. (© 1981, John W. H. Simpson.)

would encounter no resistance. In short, cosmic strings would be superconductors!

At the time of Witten's discovery, Jeremiah Ostriker, who was also at Princeton, was working on his explosion model of galaxy development. "I didn't pay much attention to strings at first because it seemed that the gravitational effects were minor," Ostriker said. "Then I received a reprint from Witten on how they could carry a high current. It was an exciting idea." But when he read the paper he realized that a critical point had been missed. If cosmic strings did, indeed, carry a high current, and if they oscillated, they would radiate. "They would act like an antenna," said Ostriker. "They would radiate away most of their energy in the form of electromagnetic waves rather than gravitational waves. And since the interaction between matter and radiation is 10^{40} times stronger than between matter and gravity, it could have dramatic consequences."

Although Witten's office was only a stone's throw away in the next building, Ostriker wrote him a letter describing his thoughts. Then when he was sure Witten had received the letter he went to visit him. And soon Ostriker, Witten, and a graduate student, Chris Thompson, were collaborating on a paper.

They soon discovered that the electromagnetic wave that was emitted by the oscillating loop was strong enough to produce giant bubbles. Space would be swept clean as the waves moved out around the strings. And chains of galaxies would form where the material collided, along the surfaces of the bubbles.

Because Ostriker had earlier proposed an explosion model of galaxy formation I asked him if he was still satisfied with it in

Bubbles around oscillating strings.

light of his new theory. "Yes," he replied. "In fact, evidence, from the IRAS satellite, is proving that I am right. There is considerable energy being put out by forming galaxies." He went on to say that this paper was not addressed to large-scale structure, only to the making of galaxies themselves. He does not believe that explosions of the type he had earlier visualized could cause large bubbles and filaments. "Cosmic string theory is needed to account for the large-scale structure," he said.

But is there any observational evidence for superconducting strings? What, in fact, would we expect to see? Because the current that flows is so large (approximately 10^{18} amperes) intense magnetic fields would be produced around the strings, and particles within the fields would glow. Cosmic strings might therefore be directly visible. And interestingly, glowing threads have been seen emanating from the core of our galaxy. Mark Morris of UCLA and Farhad Yusif-Zadeh of Columbia University discovered them in 1986.

Cosmic strings have generated considerable interest in the last few years and will likely continue to do so unless their existence is disproved. But, although some astronomers are particularly enthusiastic about them, others remain skeptical. Edward Kolb of Fermilab says, "They're interesting and need to be explained, but probably not relevant." Jim Peebles of Princeton: "They're a long shot . . . a game." And David Schramm of the University of Chicago says, "I believe that they are, at the present time, the best explanation of galaxy formation and large-scale structure of the universe. However, they are by no means proved."

SUMMARY

We now have a fairly good idea how galaxies formed. And it is quite possible that cosmic strings, if they exist, could explain the large-scale structure of the universe. Our story of creation, then, has taken us up through the creation of galaxies and large-scale structure. But for life to eventually form in the universe, we have to have heavy elements. In an earlier chapter I talked about the formation of the light elements and mentioned that the heavier ones were formed in stars. In the next chapter we will look at how they were formed.

CHAPTER 15

The "Heavy Element" Cookbook

Without heavy elements, planets such as Earth and life are not possible. But where did the heavy elements come from? In an earlier chapter we saw that for several years this was a controversial issue. Finally, though, it was resolved in favor of stars. Let us turn, then, to how they were formed. We will begin with the moderately heavy elements, those from carbon to iron. Elements beyond iron will be dealt with later in the chapter.

To understand how the elements were produced we must build a model of a star—a mathematical model. Sir Arthur Eddington gave us the equations to do this back in the 1920s, but it was many years before astronomers were able to take advantage of them. The number of calculations involved, even for an average star like our sun, is exceedingly large. So it was not until computers became available that the first models were built.

Before they could build such models astronomers also had to thoroughly understand the complicated nuclear reactions that were going on in the core of the star. This is the so-called thermonuclear furnace—the region where the energy of the star is generated. In particular, astronomers had to understand how the reactions changed as the star aged. And finally, they needed a highly efficient mathematical technique for solving the equations.

Two of the first to try their hand at a mathematical model of this type were Fred Hoyle of Cambridge and Martin Schwarzschild of Princeton University. Schwarzschild is the son of Karl

Schwarzschild, the astronomer who years earlier had obtained the first solution of Einstein's field equations. He developed an early interest in astronomy and went to Göttingen for his doctorate, graduating in 1935, about the time Hitler was coming to power. He remained in Germany for a while, but soon realized his life was in danger, so he fled to Norway. From there he came to the United States and within a short time was on the faculty of Princeton University.

Schwarzschild and Hoyle began working together in the early 1950s. It was easy enough to build a static model of a star similar to our sun, but when they tried to see what would happen as the star aged, they ran into trouble. They were able to show that it would expand and become a red giant, but that was as far as they could go.

Let us take a few moments to look in detail at how we create such a model. A star is nothing more than a sphere of hot gas—mostly hydrogen and helium, with a small amount (about 1%) of heavier elements. It is held in equilibrium by two equal and opposite forces: an inward gravitational force and an outward gas pressure. The weight of the gas in each layer of the star has to be balanced by an equal outward pressure. If not the star will either expand or contract.

The outward pressure is a result of the energy being generated in the core of the star. And this energy, in turn, is a result of nuclear reactions. It is essential, therefore, that astronomers know what type of reactions are occurring in the core. If the star is about the size of the sun the major energy cycle is referred to as the proton–proton cycle. It is a sequence of four reactions discovered by Hans Bethe in 1939. In this cycle four hydrogen atoms "fuse" to create one helium atom. Energy is generated because the mass of the hydrogen is slightly greater than the mass of the resulting helium. Matter is, in essence, being converted into energy.

If the star is slightly more massive than the sun a different cycle operates, called the CNO cycle. It also produces helium from hydrogen, but it is a hotter-burning, more efficient cycle

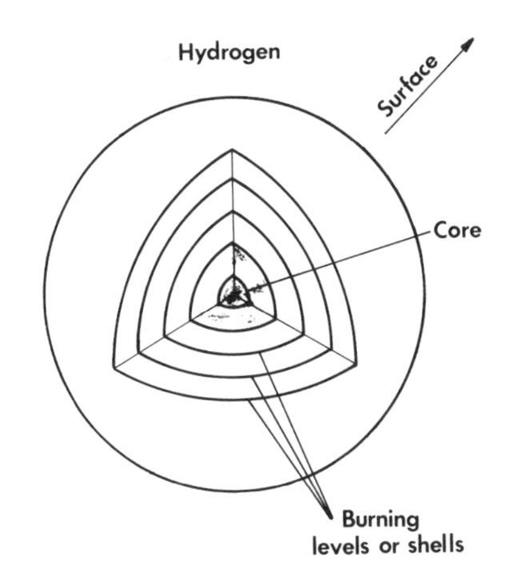

A simple schematic showing burning levels within a star.

that uses carbon (C), nitrogen (N), and oxygen (O). Once the nuclear reactions in the core are known, model building can begin. The four main properties of a star are: pressure, temperature, energy flow, and mass (weight). If we know what nuclear reactions are involved and the mass of the star we will be able to make a good estimate of these properties at its center. Also, we can either measure or determine them at its surface.

We begin by dividing the star into 100 or so layers. You can think of it as an onion, consisting of 100 shells. Our main tool is a set of four mathematical equations that tell us how the temperature, pressure, mass, and energy flow change as we go from layer to layer. With 100 layers and four equations, however, we have a total of 400 distinct equations to solve, so it is easy to see why we need a computer.

We start with the known values at the surface and begin working our way inward, one layer at a time, until we finally get to the center. In the early days, with old-fashioned computers this could take 10 or 12 hours, and there was always the danger of the computer breaking down. Furthermore, when you finally got to the center, the four quantities had to match the known ones. If they did not you had to make appropriate adjustments and start over—and most of the time this is exactly what happened. It would usually take many runs like this to get a model. And, of course, this was a model at one point in time—a static model. But the idea of building such models was to see how the star evolved. So your next step would be to let the model "age" a little, then obtain another static model, and so on until you traced the entire evolution of the star. Needless to say, the amount of work involved was horrendous. Months, and sometimes years were required. Fortunately, today we have computers that allow us to calculate static models in seconds.

Hoyle and Schwarzschild used a technique of this sort to build their models. Concentrating on stars similar to our sun, they attempted to take them beyond hydrogen burning. They wanted to find out what happened to the star when the "ash" from the hydrogen burning, namely, helium, began to burn. Incidentally, I am using the word "burning" a little loosely here. Certainly, hydrogen atoms do not burn in the usual sense. Burning in the core refers to the nuclear reactions that convert the hydrogen to helium.

Anyway, when Hoyle and Schwarzschild tried to push beyond hydrogen burning they ran into problems. They could not get the helium to burn properly. One of the major reasons, of course, was that they did not completely understand the nuclear reactions that were required to burn helium. Helium could not burn to lithium; we saw earlier that there is a gap at atomic masses 5 and 8. Somehow, when helium burned, it had to leave carbon as ash. And in 1952 Ed Salpeter showed this was possible. But the reactions were far too slow. Something was still seriously wrong.

What happened in the next few years to overcome this problem was detailed in an earlier chapter, so I will say little about it here. In short, Hoyle became convinced that an additional energy level had to exist just above helium. And he went to Kellogg Labs in California to see if it could be found. And, as we saw earlier, it was found. Shortly thereafter Hoyle got together with Fowler and the Burbidges and produced the classic paper B^2 FH, which detailed the nuclear reactions up through iron.

Let us turn now to the elements heavier than iron. The lighter elements were built up primarily by proton bombardment; the bombarding proton easily overcame the repulsive charges of the protons in the nucleus, and once it got close enough it "stuck" to the nucleus, creating a heavier element. But this was no longer possible in the case of iron. The charge barrier was so great it was virtually impossible for a proton to penetrate it. The way around this difficulty was discovered by Hans Suess and Harold Urey. They realized that element building could continue past iron if neutrons were the bombarding particles instead of protons. In looking at the plot of abundance versus atomic weight of the elements they noticed a series of double peaks in the heavy element region. On the basis of this they suggested there were two types of neutron capture: a slow (s) process and a rapid (r) process. And, as a result of these processes, elements heavier than iron could be produced. Fowler, Hoyle, and the Burbidges incorporated this result into their paper, showing that the slower neutron capture process could produce elements up to about lead, but the rapid process could go all the way to uranium.

At about the time of this discovery an important verification came. The results for the radioactivity of the debris from the first hydrogen bomb explosion were declassified. Glenn Seaborg and his colleagues soon found evidence of the heavy element californium in the debris, indicating that the neutron capture hypothesis was correct.

But even with this, astronomers were still not able to get stellar models to go beyond hydrogen burning. Things had, in fact, virtually stagnated after Hoyle and Schwarzschild's work of 1955. Hayashi in Japan, using small tabletop computers, struggled in the late 1950s to find out what happened to a star when the hydrogen was depleted. But something else was still needed—a breakthrough in calculating technique. Fortunately, it soon came. Louis Henyey of the University of California at Berkeley had, for years, been working on a new, more powerful technique. Indeed, several years earlier, he had published an early version of his method, but it was still crude and unreliable at this stage. But he had now simplified and perfected it. And at the International Astronomical Union meeting at Berkeley in 1961 he presented his new method. Virtually everyone interested in stellar models was in attendance. Among them was Martin Schwarzschild of Princeton.

Schwarzschild took extensive notes at the presentation, then went back to Princeton and began incorporating the new method into his computer program. The first run of the new program was nerve-racking. Would it allow helium to burn? It did. In fact Schwarzschild was able to use it to continue through carbon and heavier elements. Soon similar programs were being written around the world. And as computers got faster and more efficient, tremendous advances were made. Stellar model building had come of age.

ELEMENT PRODUCTION IN STARS

Now that we know that the elements can be produced in stars, let us look at how it happens. We will only consider elements up to iron for now. As I mentioned earlier, the cycles that convert hydrogen into helium are the proton–proton cycle and the CNO cycle. They require a core temperature of about 15 million degrees (the CNO cycle requires a slightly higher temperature than the proton–proton cycle). It turns out, though, that a much hotter burning cycle is required to burn helium. It is referred to as the triple alpha cycle because it involves three alpha particles (helium nuclei). The temperature in this case must be about 100 million degrees. And if we are to burn the ash of helium, namely carbon, a temperature of 600 million degrees is needed. High temperatures such as this occur only in massive stars.

Interestingly, although massive stars have much more fuel than smaller stars, they live a much shorter period of time. The reason is that they burn their fuel so much faster. A star like our sun will burn hydrogen for a total of about 10 billion years. (As our sun has been around about 5 billion years, we can expect it to remain generally the same for another 5 billion years.) A star 25 times as massive as our sun, on the other hand, will live for only a few million years. And a tiny star, say, one about a quarter as massive as our sun, will live almost 50 billion years.

Let us look at what is going on in the core of the star. When a star forms it triggers either the proton–proton or CNO cycle. The temperature in its core is about 15 million degrees at this stage. The burning hydrogen produces helium which, being heavier than hydrogen, falls to the center of the star. And as it builds up, hydrogen burning takes place in a shell around it. The helium core at the center continues to get larger for a while, then it begins to contract. And as it contracts it gets hotter.

At this point everything is still under control. The pressure of the gas depends on its temperature, and as a result there is a "thermostat" that keeps the star under control. If the reactions start to produce too much energy the temperature of the gas increases, and this in turn raises the pressure. The increased pressure causes the star to expand which then cools it, slowing the nuclear reactions.

This keeps the hydrogen burning under control. But when the helium at the center begins to burn we have a different story. The helium continues to contract until it reaches a temperature of 100 million degrees. It then ignites. But in a medium-sized star such as the sun the helium has no "thermostat." Pressure and temperature are not related, and an increase in temperature does not cause an increase in pressure. There are therefore no controls when ignition occurs. It's almost as if the fuse to a keg of dynamite has been lit. The nuclear reactions just go faster and faster as the temperature shoots up. Soon they are out of control and an explosion, referred to as the helium flash, occurs. (Incidentally, we refer to the material in which this happens as "degenerate.")

The explosion blows the inner part of the star completely apart. The hydrogen burning ring is obliterated, and the helium core shattered. Strangely, though, despite the magnitude of the explosion, no one outside the star would even notice it. At this stage the envelope of gas around the core is so extended it completely conceals the explosion. High temperatures have driven it off into space. Our sun at this stage would have its outer layer out near the orbit of Mars.

But eventually we would notice a difference. All burning in the core has ceased so the star will soon begin to grow dimmer. Over a period of years it will grow considerably dimmer, but as the helium falls back into place burning will begin again—this time peacefully. The hydrogen will burn in a ring around the helium core, and the helium will burn at its center. And, as a result, the star will brighten again, eventually achieving the brightness it had before the explosion.

This helium flash does not occur in all stars. If the star is less than about 0.4 times the sun's mass, helium will never burn because the core will never get hot enough. If, on the other hand, the star has a mass greater than about 3 solar masses the helium will be ignited before the star becomes degenerate, so nothing will get out of control.

Now, just as hydrogen burning leaves ash, so too does helium burning. The ash in this case is carbon and oxygen. And again they are heavier than helium so they accumulate at the center of the star. And soon we have a shell of burning helium around a core of carbon and oxygen. Out farther, of course, there is a shell of burning hydrogen.

The carbon–oxygen core grows, then contracts and heats. If the temperature at its center gets to 600 million degrees the carbon in it will ignite. And again we will have an explosion. This explosion, though, is much more powerful than the helium flash. But again it does not occur in all stars. Stars less than 3 solar masses do not generate core temperatures as high as 600 million degrees. And stars with greater than 9 solar masses burn before they develop a degenerate core. (This is, of course, outside the range of our sun, so we can say that carbon detonation will never occur in it. It will, however, develop a carbon–oxygen core.)

Getting back to the explosion we ask: What would it be like? It would certainly be visible outside the star this time. In fact, the entire star would be blown apart (except for a small region at the core). We refer to such explosions as supernovae. We will talk about them in detail later.

If the star is sufficiently large—greater than about 9 solar masses—there is no explosion. The star will just ignite the carbon and burn it peacefully. The ash in this case is neon.

This sequence of steps will continue through oxygen, magnesium, silicon, and iron. And at this stage the star will be burning on numerous shells, one within the other. This is, indeed, how the elements of the universe, up to iron, were generated. The next problem is, of course, how these elements built planets and solar systems and even life. To answer this we have to look at the death of stars.

DEATH OF A STAR

We have seen that small stars, those with a mass less than 3 solar masses, live long, generally uneventful lives. They burn hydrogen, then helium, but they cannot ignite the carbon in their core. They do, however, undergo one last dramatic event: late in their lives the reactions within them get out of control. This causes a sudden heating of their surrounding envelope. And soon the entire outer shell of the star expands into space, leaving a small hot star. Its surface temperature is extremely high, but its thermonuclear furnace has gone out and gravity soon starts to overcome it. As the shell of the cooling gas moves outward, the star cools. From a distance the tiny star looks as if it has a smoke ring around it. We refer to such objects as planetary nebulae.

Over millions of years the central star shrinks and cools. But as it shrinks pressure caused by the electrons in the gas builds up and eventually halts the collapse. This happens when it gets to a size slightly larger than the Earth. It is now a white dwarf. The material from such an object is so heavy that if you brought a teaspoon of it to Earth it would weigh 50 tons.

Subrahmanyan Chandrasekhar showed many years ago, however, that electron pressure can only hold up a star if it has a mass less than about 1.4 solar masses (called the Chandrasekhar limit). What happens if the mass is greater than this? The star simply burns carbon, which again leaves ash. And if the star has a mass greater than about 9 solar masses this will continue all the way to silicon. The ash from silicon is iron. But iron is different from the elements we have talked about so far: it cannot burn by thermonuclear reactions. In the reactions we have talked about so far energy is given off because the neutrons and protons were packed closer and closer together with each successive cycle. But in iron they are packed as closely as they can be; iron is the most stable of the elements and cannot be compressed. Fusion is no longer possible.

Iron cannot burn, but if the star is to remain extended its core must be held up against the tremendous pressure of gravity. And for a while it is—by electron pressure. But as more and more silicon is burned the iron core grows more massive. Finally it exceeds the Chandrasekhar limit and begins to collapse.

SUPERNOVA!

The last stage of the buildup occurs rapidly. Each successive cycle takes a shorter and shorter time. It took millions of years to create a helium core hot enough to burn, but it takes only a day to build up the iron core.

Although the star is huge at this stage, perhaps bigger than the orbit of Jupiter, the iron core deep inside it is smaller than the Earth. As the collapse begins the core heats rapidly, then it begins to crumble. Iron nuclei break up into smaller nuclei, then individual particles. Free protons and electrons start to combine, forming neutrons and in the process giving off neutrinos. These neutrinos leave the star rapidly, carrying huge amounts of energy with them. In addition, because the core is being supported by electrons at this stage, the loss of so many of them greatly accelerates the infalling matter.

The entire collapse takes only milliseconds. The density of the core continues to increase; finally the core becomes so dense it stops the neutrinos from escaping. The inner part of the core then starts to stabilize. It begins to act like a huge solid—what astronomers call a homologous sphere—retaining its shape as it collapses. But its density continues to increase until it reaches the density of nuclear matter. It is now like a gigantic nucleus, and therefore incompressible.

As the center of the core grinds to an abrupt halt sound waves echo back through it. Then the outer infalling layers crash into the surface of the gigantic nucleus with a tremendous force. The incredibly rigid nucleus stops them, but not abruptly. There is a small amount of compression. Then, like a hard rubber ball, the core rebounds, sending a powerful shock wave speeding outwards. The star explodes in a dazzling display of power. The resulting brightness is greater than that of a billion suns. Radiation, neutrinos, and gravitational waves streak out from deep within the star's bosom. Stellar debris is flung out in all directions at speeds close to that of light. The supernova has begun.

THE HEAVY ELEMENTS

So far I have talked mostly about the elements lighter than iron. Where did the elements heavier than iron come from? The answer to this was given in the paper by Fowler, Hoyle, and the Burbidges. I mentioned that they discussed the possibility of neutron capture via two processes—the s and r processes—and it is these processes that give the heavy elements.

Where do they occur? As you might expect, in the super-

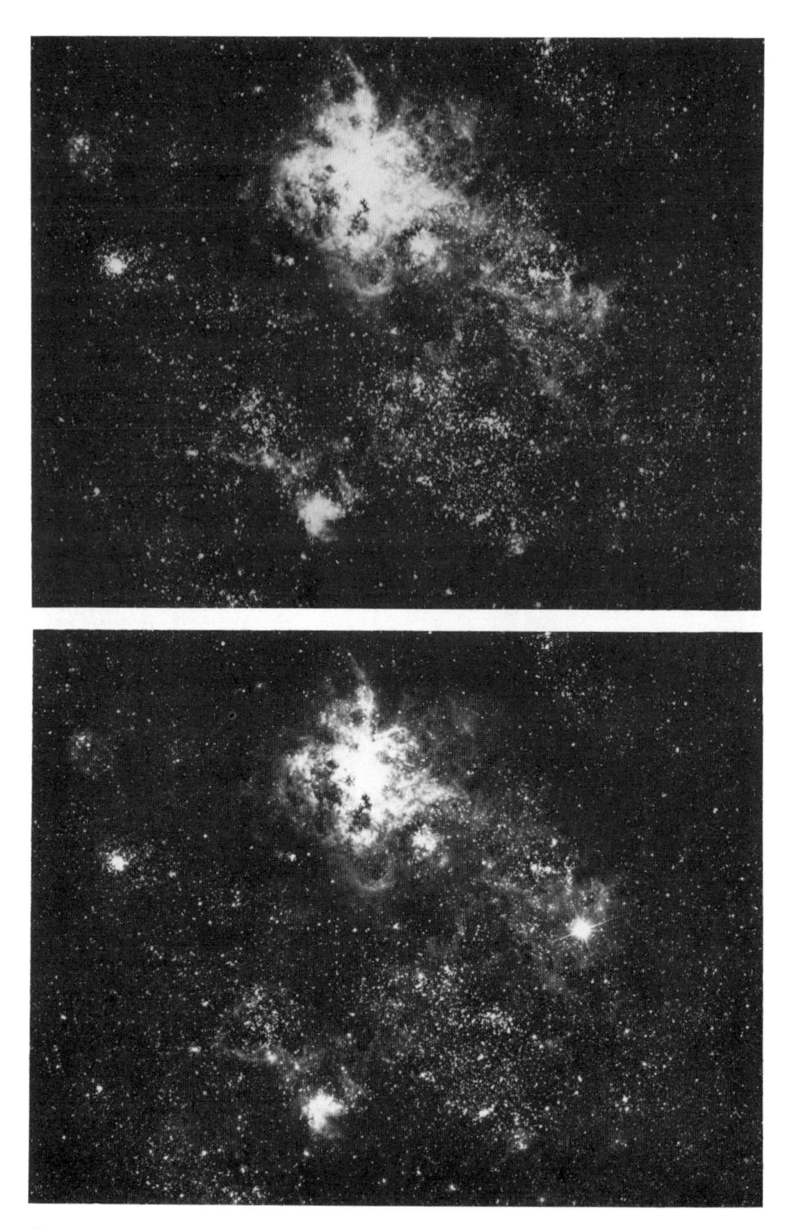

Supernova 1987A. Before (upper) and after (lower). (Courtesy National Optical Astronomy Observatories.)

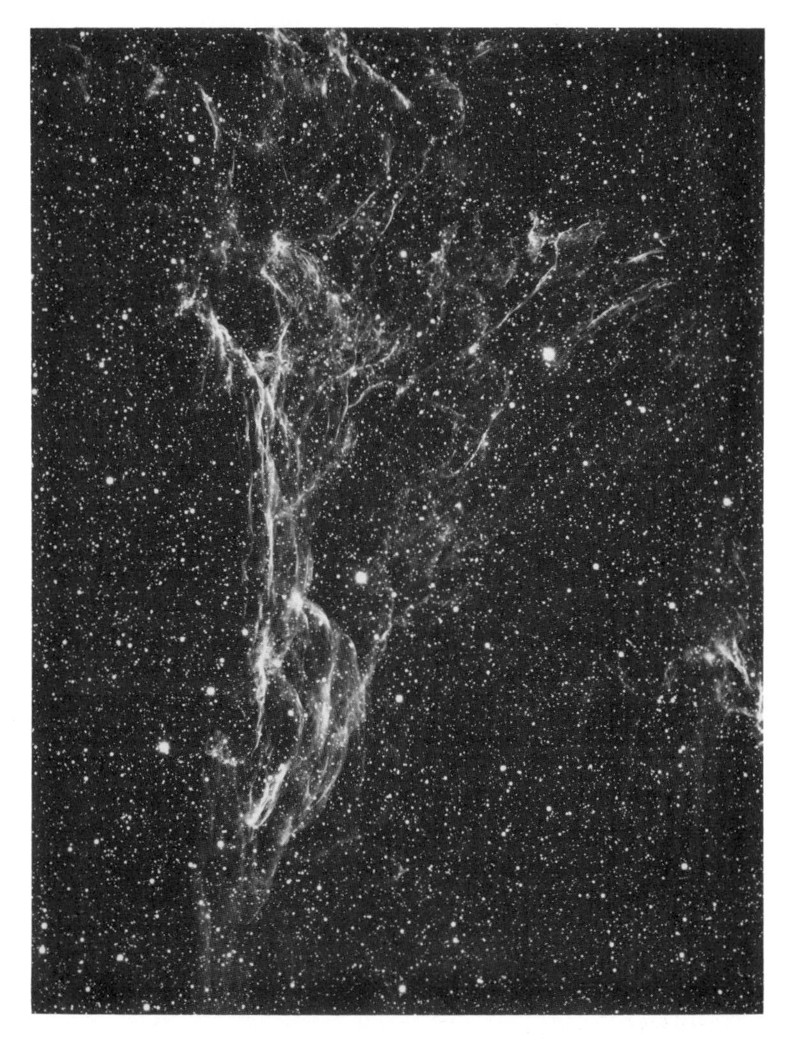

Remnant of a supernova explosion. (Courtesy National Optical Astronomy Observatories.)

nova explosion. This means that everything beyond iron is produced in the very short period of time during which the supernova explosion occurs. In short, all elements up to iron (with the exception of a few of the lighter ones) are cooked deep in the interior of stars. And beyond iron they are generated in the supernova explosion.

If you think about it, you realize that nature seems to have been put together in exactly the right way. Most of the elements are made in extremely massive stars (greater than 10 solar masses), but these are the stars that live the shortest time. They are also the stars that explode and distribute the heavy elements into space so that solar systems such as ours can form. If most of the heavy elements were made in small stars that lived, say, 50 billion years and never exploded there would obviously be problems.

It may be difficult to believe, but literally every atom in your body was at one time produced inside a giant star somewhere in the universe. In fact, many of them were produced in supernova explosions.

CHAPTER 16

The Emergence of Life

The stage was now set. Once heavy elements were available planetary systems such as ours could form. And in them life could develop and evolve. How did it happen? There are, of course, many uncertainties, but scientists feel they now have a good idea. Let us start with the formation of the solar system and Earth.

About five billion years ago there was a huge gas cloud, about twice as massive as our sun. It was the remnant of a giant star that lived for a few million years and exploded. This cloud, which we now refer to as the solar nebula, was made up mostly of hydrogen and helium, with a dash of heavier elements (about 1%). As it spun it contracted and heated.

The outer regions of the cloud remained relatively cool as its center heated, and this caused a temperature gradient throughout it. In time elements began to condense out of the cloud as grains, but because the temperature was so much higher closer to its center, different elements condensed there as compared to its outer regions. Heavy elements such as iron and nickel condensed close to the center whereas lighter materials such as methane and ammonia condensed farther out. This is why Mercury, Venus, and Earth have large iron–nickel cores, and Jupiter and Saturn have considerable methane and ammonia.

The grains eventually moved to the midplane of the nebula and formed a huge sheet—like a giant version of Saturn's rings. As they whirled they struck one another and coalesced until small rocks were formed. This aggregation continued until the rocks were a few miles across. At this stage they are referred to as planetesimals. These planetesimals continued to collide and coalesce until finally they became protoplanets. It is important to remember that, at this stage, they were still immersed in a heavy fog.

While the protoplanets were forming, the mass at the center of the nebula—the protosun—was getting denser and its temperature was increasing. Finally its central temperature reached 15 million degrees and nuclear reactions were triggered. And with the triggering came a sudden explosion—the solar gale. This gale rushed out into the solar system sweeping away the hydrogen and helium, leaving the inner planets without an atmosphere. It was not strong enough, though, to blow the fog from the giants farther out and as a result it gradually condensed onto them. This is why we now have small terrestrial-like inner planets and large gaseous giants farther out.

FORMATION OF THE ATMOSPHERE

With the solar gale the first atmosphere of the Earth (mostly hydrogen and helium) was blown off into space leaving a barren, desolate surface. But beneath this surface something was stirring. Radioactive materials were releasing energy, heating the rocks around them, until finally they were molten.

Earthquakes then shook the surface, creating small fissures. Soon the molten lava began trickling upward, forcing its way through the cracks. At times the pressure became so great that it ruptured the crust and exploded with devastating force, throwing clouds of gas and dust miles above the surface.

Although the molten lava sometimes exploded when it broke through the surface, most of the time it just flooded out and down the sides of an ever-growing volcanic dome. For millions of years the volcanoes belched gas and water vapor, eventually creating a new atmosphere. Most of this atmosphere was carbon dioxide, but there was also nitrogen and water vapor. And while the volcanoes exploded, water condensed onto the surface, and oceans began to form in the low-lying areas. The carbon dioxide from the volcanoes reacted with the material in the oceans creating limestone and quartz which eventually went into sea shells and rocks. Without these reactions the Earth's atmosphere would now be mostly carbon dioxide, as are the atmospheres of Venus and Mars. There was no ocean when the volcanoes on these planets expelled their carbon dioxide, and as a result it stayed in the atmosphere, and is still there today.

The oceans were huge, but generally shallow. And they were hot—heated by the radioactivity in the surface beneath them and by the volcanoes that rose within them. But gradually they cooled. In the newly formed atmosphere of carbon dioxide, nitrogen, and water vapor above them chemical reactions were occurring. The carbon dioxide was rapidly disappearing, but the reactions were producing methane and ammonia. Soon the atmosphere consisted mostly of methane, ammonia, nitrogen, and water vapor—an atmosphere we now refer to as the primitive atmosphere. It was what we call a "reducing" atmosphere.

We will see later that such an atmosphere was critical to life, for it is only in such a mixture that life can form. The first forms of life, some of it algae, no doubt appeared in the oceans. The algae were important in the evolution of the atmosphere. They converted carbon dioxide into plant material and oxygen. And as a result, free oxygen soon appeared in the atmosphere. In time it accumulated and became more common than methane and ammonia. The atmosphere, in effect, changed from a reducing one to an oxygenating one. Once there was oxygen, an ozone layer formed and the life forms below were shielded from the deadly ultraviolet rays of the sun.

Ultraviolet light, incidentally, is one of the major reasons that Mars does not appear to have any life today. It does not have an ozone layer, and any life forms on its surface would have to be shielded if they were to survive.

WHAT IS LIFE?

Before we can talk about how life formed in the atmosphere we must look at what life is. It is, indeed, difficult to define exactly what it is, but most scientists will agree that the main thing that characterizes it is reproduction. Even with this rather simple definition, though, we have to be careful. Some things that are not alive do reproduce themselves. Crystals and viruses are two examples (although there is some question about viruses).

The basic unit of life is the cell. It is the cell that reproduces itself, making duplicates on a regular schedule. Inside the cell we find the basic molecules of life. At first glance they appear complicated because they are extremely long. But on closer inspection they are seen to be relatively simple—built from a small number of basic units. The master molecule, the molecule that controls everything that goes on in the cell, is deoxyribonucleic acid, or DNA for short. It works in conjunction with a similar molecule called RNA.

The structure of DNA was determined by Francis Crick and James Watson at Cambridge University in 1953. They showed that it is helical. There are, in fact, two helices made up of sugar and phosphate that are wound around one another giving the appearance of a spiral staircase. Holding these helices together are the four bases adenine (A), thymine (T), guanine (G), and cytosine (C). They occur in pairs across the helices. Critical to the function of DNA is the fact that A bonds only with T, and C only with G. Furthermore, the bonds are weak, so it is relatively easy to separate the pairs. This means the DNA molecule can break the bonds along its center and unwind. And when it does, it exposes a sequence of bases (e.g., ATTGCC. . .). This sequence is a code—the code of life. It codes for everything that is needed by the cell. Its preservation is critical to the existence of the cell, and it is therefore well protected. Only at replication time is it exposed.

RNA is similar to DNA in that it also has a double helical

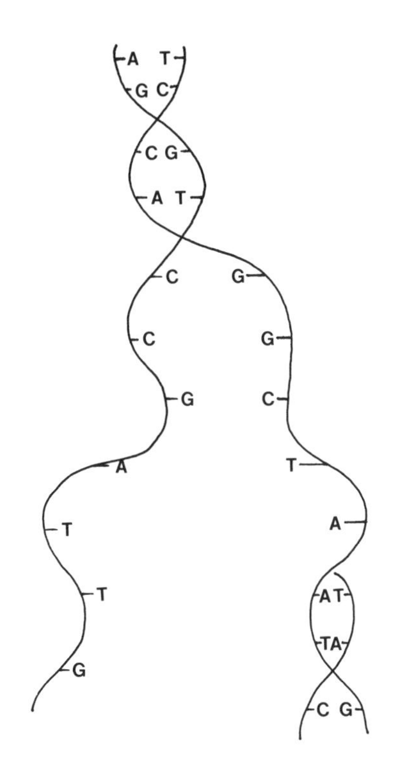

A DNA molecule in the process of unwinding. Note the code.

structure with a sequence of bases holding the two side strands together. The only difference is that thymine (T) is replaced by uracil (U), and the deoxyribose sugar in the side strands is replaced by ribose sugar. The role that RNA plays, however, is quite different. It can be thought of as the "worker" molecule of the cell. There are, in fact, several different kinds of RNA that play different roles. But, in general, their major function is to produce the molecules necessary to keep the cell running.

The other important molecule is protein. Like DNA and RNA it is a long chain made up of simple units: amino acids. There are about twenty of these acids. Proteins are the enzymes

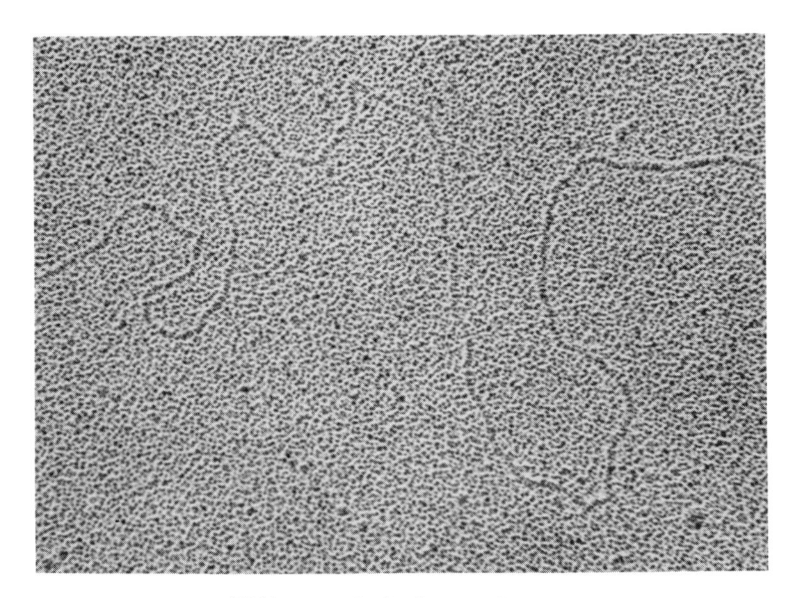

DNA as seen in the electron microscope.

that fuel the cell. DNA codes for them; RNA picks up the code and makes them.

These, then, are the basic molecules of life. The important question now is: Could they be produced naturally in the atmosphere of the early Earth? More exactly, could the components of these molecules be produced, and could these components later come together to make the long molecules required? We will look at that in the next section.

EARLY EXPERIMENTS

For years it was generally accepted that life began from nonlife. Spontaneous generation it was called. Everyone knew that worms formed in mud, and maggots soon appeared in the carcasses of dead animals. In time, though, doubt began to arise, and by the mid-1800s there was considerable controversy over spontaneous generation. The French Academy of Science decided in the early 1860s to offer a prize to anyone who could prove once and for all whether or not life could form from nonlife. The French chemist Louis Pasteur took up the challenge, and in a classic series of experiments showed that spontaneous generation was impossible: life could not form from nonlife. And for years his result was accepted.

But Pasteur had shown only that spontaneous generation could not occur in our atmosphere. What about a different atmosphere? And what about a time span of millions of years? Pasteur had only allowed a year for his experiment. In 1924 the British biologist J. B. Haldane considered the possibility of a different atmosphere. Looking at the large amount of carbon deposited in the ground in the form of coal, he arrived at the conclusion that there must have been a lot of carbon dioxide in the atmosphere at one time. He also concluded that there was no oxygen in the early atmosphere of Earth, and therefore lightning, volcano eruptions, and ultraviolet rays from the sun would act on the gases of the atmosphere to produce biochemical compounds. These compounds would then be washed into the oceans until they eventually formed a "primordial soup." According to Haldane, life formed in this soup. Similar ideas were put forward about the same time in Russia by Aleksandr Oparin. For the most part, though, both men were ignored. One of the reasons, no doubt, was that the idea offended many people; they seemed to be atheistic. And therefore for years they lay undeveloped.

The first experiment to check on the hypothesis was not performed until 1953. Stanley Miller, a graduate student at the University of Chicago, went to a lecture on the origin of life given by Harold Urey, who was then also at the University of Chicago. During the lecture Urey mentioned that many people had tried to create life from nonlife but had failed. The reason for their failure, he said, was that they were using the wrong atmosphere. The early Earth had a reducing atmosphere of methane and ammonia, and it was in this atmosphere that life arose.

After the lecture Miller went to Urey and asked if he could do this experiment as a thesis project. Urey agreed, and Miller soon began to build an apparatus—a basically simple apparatus. He filled a flask with purified and sterilized water, then circulated an "atmosphere" of methane, ammonia, and hydrogen above it. This was a simulation of the atmosphere of the early Earth. Of course an energy source would also be needed. On the early Earth it would probably have been ultraviolet light from the sun, or perhaps lightning. Miller therefore used an electric arc. He allowed the atmosphere to circulate through his apparatus with the arc discharging for about a week. Then, examining the products in the water and on the inner surface of the flask, he found organic molecules. Haldane and Oparin were right. In particular, Miller found several types of amino acids—the building blocks of protein. This meant that protein could be produced naturally in the atmosphere of the early Earth.

But what about the other molecules—DNA and RNA? Could their components also be generated? In 1961 Juan Oro, working at the University of Houston, showed that adenine, one of the bases of DNA, could be produced in a similar way. Scientists quickly jumped on the bandwagon. The other bases proved more difficult, but by adding catalysts (agents that speed reactions but do not take part in them) that should have been present on the early Earth, and a variety of energy sources, they were eventually obtained, as were the sugars and phosphates of the side strands. But there was still another important step. Would the components form a DNA molecule when mixed together? The experiment was performed and, as hoped, DNA did form when appropriate catalysts were used. Scientists have also shown that mixtures of amino acids can produce simple proteins.

THE EMERGENCE OF LIFE ON EARTH

Now that we have seen that the basic molecules of life can form in the lab, let us consider how it likely happened on the early Earth. About four billion years ago conditions were ideal: the atmosphere was reducing and there were numerous energy sources available.

The surface of Earth was still in upheaval. It was being hit continuously by meteorites; earthquakes were common, volcanic explosions occurred daily, spreading debris and ash over a wide area. In addition there were tremendous storms, with torrential rains and almost continuous lightning. All of these were potential energy sources, but the most important source, by far, was ultraviolet light from the sun. Acting on the reducing atmosphere it would have produced organic molecules—some of them the components of DNA and protein.

But these molecules, once produced, needed protection from the sun's strong ultraviolet rays. Otherwise they would soon be broken down. How did they get protection? The most likely possibility is that soon after they were formed they were washed into the oceans in the torrential downpours that occurred. Once in the oceans they would be safe. Indeed, they could react and form further organic molecules.

Unfortunately, there was now a serious problem. Large, chainlike molecules like DNA and protein (called polymers) cannot form in a dilute solution of water. Water prevents bonds from forming. The components had to become concentrated, and to do this they somehow had to leave the oceans. But there was a catch-22: if they did they would soon die under the strong ultraviolet radiation. Fortunately the life forms in the water helped themselves through a process called photosynthesis. As a result of this process oxygen was produced and soon an ozone layer formed that shielded them from the ultraviolet light.

Life forms could now leave the oceans. There are several ways this could have occurred. Tides no doubt played an important role. Evaporation was also no doubt important. As a pond evaporated, for example, the material in it became more and more concentrated, and this was, of course, what was needed for polymers to form. Freezing may also have helped. The water may have frozen, leaving the rich broth of nutrients behind. There are indications also that clay played an important role. It is difficult to say which of these was most important, but once it happened—once the primordial soup got thick enough—long molecules began to form. And the first DNA, RNA, and protein molecules appeared on the surface. Of course, it is still a long ways from here to intelligence.

THE RISE TO INTELLIGENCE

The part of the story that I have told so far, namely the formation of the first molecules of life, although most distant from us in time, is the best understood part. From here on most of what we now accept is based to a large degree on speculation. Nevertheless, it is a reasonable picture, and most scientists accept it.

From the first life molecules came the first cells. The first organisms were, no doubt, single-celled creatures. In time—millions of years—multicellular organisms developed. And in more millions, perhaps billions of years, the first vertebrates (creatures with a backbone) appeared. It took a long time for the first life forms to appear, perhaps as long as one and a half billion years. But the rise to intelligence took even longer—another three billion years.

Do we know for sure that intelligent life would form if the conditions were right? Certainly we cannot say for sure, but it does seem likely. We have to admit, though, that many conditions are critical. Indeed, most scientists believe that if we wiped the slate clean and started over again, the life forms that evolved would be quite different—particularly the intelligent forms (assuming they evolved). It is indeed strange that there have been several million different species on Earth, yet only one has evolved to higher intelligence.

LIFE ELSEWHERE IN THE UNIVERSE

Our discussion of the evolution of life would be incomplete if we did not discuss the possibility of intelligence beyond the

Earth. We have journeyed to several of the planets in our system and know that the probability of life there is low. The best candidate is Mars. As you likely know, it was checked during the Viking visit and nothing was found. Beyond our solar system, however, the outlook is much brighter.

If life is to exist elsewhere there must, of course, be planets to sustain it. Do we have any evidence that there are planets around nearly stars? Let us consider this first.

So far we do not have absolute proof of extrasolar planets, but several searches are now in progress and it is quite possible that a breakthrough in this area could occur anytime. There have, in fact, been a number of important recent developments, one of which was the discovery of a cloud of debris around the bright star Vega. Most of this debris appears to be small, but objects as large as planets cannot be ruled out.

The cloud around Vega was an important discovery but it was not the first time evidence for an extrasolar planet was found. That honor belongs to Peter van de Kamp, now of the University of Amsterdam. In 1945 he and several colleagues at Sproul Observatory (Swathmore College) began a search of several nearby stars. They concentrated on a small red one about 5 light-years away called Barnard's star. Over the years they recorded its position on photographic plates and carefully measured its path through the sky. And indeed they found that its path had a periodic wobble, indicating that it was accompanied by a dark object—perhaps a planet. (The two objects revolve around their common center of gravity, and as they move across the sky the visible object appears to wobble.)

van de Kamp's discovery was heralded as a significant breakthrough. But when another astronomer, George Gatewood, then of Allegheny Observatory (University of Pittsburgh), tried to verify the result he could not. This caused considerable controversy, and when it was discovered that the Sproul telescope had been moved during the observations, the controversy heightened. But van de Kamp continued his observations and finally published strong evidence for a periodic variation. According to his calculations there are two planets about the size of

The Coma Berenices cluster of galaxies. Each of these galaxies contains hundreds of billions of stars. (Courtesy National Optical Astronomy Observatories.)

Jupiter associated with Barnard's star. The controversy is not yet completely resolved, but scientists now generally accept van de Kamp's result.

Most astronomers are optimistic that there is life, perhaps intelligent life, elsewhere in the universe. The major reason is the number of stars—200 million of them in our galaxy alone. Also, it seems reasonable that the processes that gave rise to life on Earth gave rise to life elsewhere. We have no reason to believe this would not be the case—assuming the conditions were the same. Indeed, if this was not the case it would make us special and most scientists are unwilling to accept this.

There are, however, a small number of scientists who do not agree. Frank Tipler of Tulane University is one of them. He is convinced that we are the only intelligent life in our galaxy. He bases this conclusion on the high probability that if there were civilizations beyond Earth, they would soon (astronomically speaking) colonize our entire galaxy. He points out that we are, even now, capable of sending probes to the stars. We have, in fact, sent out two probes: Pioneer and Voyager. Granted, it will take a long time before they encounter any stars, but when considered in the proper perspective this is not really that long. If, for example, they were headed for our nearest star, Alpha Centauri, they would take about 80,000 years to reach it. On the other hand, in the not-too-distant future we should be able to send probes at much greater speeds, say 10% the speed of light. In this case it would take only 45 years to reach Alpha Centauri.

Michael Papagiannis of the University of Boston is also convinced that there should be a "wave of colonization" throughout our galaxy. He believes that advanced civilizations would likely use large space satellites, capable of housing hundreds, perhaps

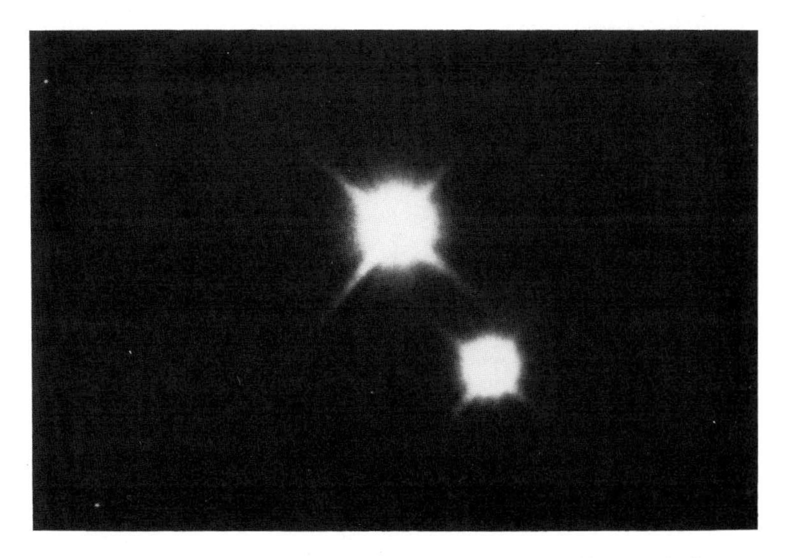

Alpha Centauri. (Courtesy National Optical Astronomy Observatories.)

thousands of people. If these satellites traveled at, say, 5% the speed of light they could colonize our entire galaxy in a relatively short period of time. Allowing 500 years for a trip of 10 light-years and 500 years for the establishment of a colony he shows that it would take only 10 million years to populate our galaxy. Because our galaxy is at least 15 billion years old, this is an enigma.

If there were a million civilizations in our galaxy, as the optimists believe, says Papagiannis, then at least some of them should have sent out a wave. Because we see no evidence of it he believes we are the only civilization in our galaxy.

Who is correct? It is difficult, of course, at this stage, to say. With the large number of stars available and the assumption that we are not extraordinary in any way, it seems unreasonable that we would be the only advanced civilization in our galaxy. On the other hand, the arguments on the other side are persuasive.

CHAPTER 17

Epilogue

Our story is now complete: we have traced the universe from its beginning to the formation of life. The picture we have presented, most astronomers would agree, is fairly accurate. Yet many questions remain unanswered. Some of our ideas, for example the explanation of the lack of antimatter in the universe, are based on the existence of a grand unified theory. There are several versions of this theory, and we are not yet certain which, if any, is correct. Indeed, it is now quite possible that all grand unified theories will be superseded by a superstring theory, and superstring theories may give us a quite different picture of creation. Without a doubt, though, if we are to thoroughly understand creation we must first completely understand the particles and forces of the universe. And we can get this understanding only if new, larger particle accelerators such as the superconducting supercollider (SSC) are built. It is also important that larger telescopes be built, and some of them be put into space.

But what will happen if new tests show us that some of our ideas are wrong? Obviously we will have to start over and look for a better theory. It is, in fact, possible that this will happen. Inflation theory may eventually be shown to be incorrect. It is even possible that the big bang theory is wrong. Only time will tell.

Turning now to the other end of the story, there is still the question of what will eventually happen to the universe. For an answer we must look again at Friedmann's theory. It tells us that the universe will either expand forever, or stop expanding and collapse back on itself, depending on its average density. If the average density is over a certain critical amount, the mutual gravitational pull of the matter will eventually stop the expansion. If not, the universe will expand indefinitely. We are still not certain which of these will happen, but as we saw earlier, the density appears to be very close to the dividing line.

Before I end the book I feel I should address a question that some of you will no doubt be wondering about: Why does a book on the creation of the universe have no mention of God? Scientists do, indeed, rarely mention God when they talk about creation. Furthermore, they are sometimes accused of trying to do away with the need for a God by attempting to explain creation in scientific terms. And it is true: scientists would prefer a purely scientific explanation of the beginning of the universe. That is not to say, though, that all scientists are atheists (few are). Furthermore, there is no fear that scientists will ever eliminate the need for a God. If we look back at the early universe we see that regardless of how far things are pushed—even if we were someday able to explain creation itself in an entirely satisfactory scientific way—there is still something that is unexplained. Creation depends on the basic laws of nature—without them it would not be possible. Who created these laws? There is no question but that a God will always be needed.

Glossary

Aether A hypothetical substance believed at one time to permeate the universe. Needed to propagate waves.

Anisotropy Different in different directions. Not isotropic.

Antigalaxy A galaxy made of antimatter.

Antiparticle Corresponding to every type of particle there is an antiparticle. When an antiparticle and a particle meet they annihilate one another with the release of energy.

Antistar A star made of antimatter.

Atomic mass The mass of an atom in (approximate) units of the mass of the hydrogen atom.

Baryon A heavy particle. Made up of three quarks.

Baryon number A quantum number. Equal to 1 for baryons, -1 for antibaryons, and 0 for other particles.

Beta decay Decay with the release of an electron. A neutron beta decays to an electron, a proton, and an antineutrino.

Blueshift A shift of the spectral lines toward the blue end of the spectrum. Indicates an approaching object.

Boson A particle with integral spin.

Causally connected A "connection" satisfying the causality principle.

Centrifugal force A fictitious outward acting force in circular motion.

Cepheid variable A star that changes its brightness periodically. Named for the brightest of the group, δ Cephei.

Closed universe A universe in which the recession of the galaxies eventually stops. Positively curved.
Cluster A group of stars, or galaxies.
Color Quality of quarks, like electric charge. The color force is the force of attraction between quarks.
Conservation principle A law that states that a quantity does not change in a physical process.
Constellation A group of stars that appear to be close to one another in the sky.
Cosmic background radiation Radiation released early in the history of the universe that now fills it. Currently has a temperature of 3 K.
Cosmic ray Particle from space that strikes the atmosphere of Earth.
Cosmological constant A constant term added by Einstein to his field equation to make the universe stable.
Cosmology A study of the structure and evolution of the universe.
CP conservation Says that every law of nature must be the same if we replace particles by antiparticles and observe in the mirror.
Critical density Density at which the universe is flat. Dividing line between open and closed universes.
Cross section The "target area" an incoming particle sees. A measure of the probability that a reaction will occur.
Dark matter Matter that appears to be missing. Matter that is not observed.
Deuterium A heavy form of hydrogen. Nucleus contains one proton and one neutron.
Domain wall A wall-like defect left after a "freezing" in the early universe.
Doppler effect The apparent change in wavelength of light due to relative motion between source and observer.
Dwarf galaxy A small galaxy. Considerably smaller than the Milky Way.
Electron pair An electron–positron combination.

Electron volt (eV) The amount of energy acquired as an electron moves through a potential difference of one volt.

Electrostatic force Force associated with an electric charge.

Era An interval of time.

Event horizon Surface of a black hole.

False vacuum An energy state of the early universe. Not the true vacuum.

Fermion A particle with a spin 1/2. A "matter" particle.

Fluctuation A small change in density.

Gamma ray The most energetic form of electromagnetic radiation.

General relativity A theory of gravity formulated by Einstein in 1915.

GeV One billion electron volts.

Globular cluster A group of a few hundred thousand stars (sometimes a few million).

Gluon Exchange particle of the strong interactions.

Grand unified theory (GUT) A theory that attempts to unify the electromagnetic, weak, and nuclear forces.

Higgs field An energy field that presumably existed in the very early universe (and still exists).

Inflation theory A theory that suggests that a sudden increase in the expansion rate of the universe occurred shortly after the big bang.

Interstellar cloud A gas cloud in the region between stars.

Irregular variable A variable star that does not have a regular period.

Isotropic The same in all directions.

Laws of physics Basic laws such as the conservation of energy and momentum.

Lepton The "light" particle of the universe. The electron, muon, tau, and their neutrinos.

Light-year A measure of distance in astronomy. The distance light travels in one year.

Local group The group of galaxies to which the Milky Way belongs.

Longitudinal wave A wave that vibrates in the direction that it is traveling.
Long-ranged force A force that extends over a long distance.
Meson A medium-weight particle. Made up of a quark and an antiquark.
MeV One million electron volts.
Monopole A heavy particle with either a south or north magnetic pole, but not both.
Muon A heavy "cousin" to the electron.
Naked singularity A singularity not "clothed" in an event horizon.
Nebula An older term for galaxy. Now refers to a region of gaseous material in space.
Neutrino A particle that is believed to be massless, that is, electrically neutral, and experiences only weak interactions.
Nova An exploding star.
Nucleosynthesis The process of generating nuclei in stars.
Open universe A negatively curved universe. Will expand forever.
Parity Pertaining to whether the mirror reflection of a process is the same.
Particle accelerator A machine that accelerates particles to high velocities (energies).
Particle pair A particle–antiparticle combination.
Phase change A change from one "state" to another, e.g., change from water to ice.
Photosynthesis A process in which green plants convert solar energy for their growth and development.
Positively curved universe A closed universe. Will eventually stop expanding and collapse back on itself.
Positron Antiparticle of the electron.
Primeval galaxy First galaxies to form in the universe.
Quantize To make into a quantum theory.
Quantum mechanics A branch of physics dealing with the structure and behavior of atoms and their interaction with light.

Quark An elementary particle. Comes in six types and three colors.
Quasar Energetic objects in the outer regions of the universe.
Radial velocity Velocity along the line of sight.
Radiation Photons; electromagnetic energy.
Radioactivity The process by which certain nuclei decay and emit particles.
Radio galaxy An active galaxy. Emits radio waves.
Radio telescope Telescope designed to pick up radio waves from the sky.
Reaction rate Speed with which a nuclear reaction proceeds.
Redshift A shift of spectral lines in the direction of the red end of the spectrum. Indicates recession.
Referees Scientists who judge the merit of scientific papers.
Short-ranged force A force that exists only over a small distance (e.g., over the size of the nucleus).
Singularity A point of infinite density. A point where the laws of physics break down.
Spectral line A line that appears when the light from a star or other object is passed through a spectroscope.
Steady state theory A theory put forward in the late 1940s that assumes that the universe was always the same and will continue to be the same into the infinite future.
Stellar synthesis The production of nuclei and so on in stars.
Strong interactions Interactions associated with the strong or color force.
Supercluster A cluster of galactic clusters.
Supergravity An extension of general relativity. A theory of gravity.
Supersymmetry A symmetry in which bosons and fermions are two states of the same particle.
Symmetric (referring to universe) Having equal amounts of matter and antimatter.
Transverse wave A wave that vibrates in a direction perpendicular to its direction of travel.
Tritium A heavy form of hydrogen.

Vacuum fluctuation Generation of a pair out of the vacuum.
Void A large region in space that contains little or no matter.
W particle Exchange particle of the weak interactions.
X particle An extremely heavy particle that may have existed in the early universe.

Further Reading

Asimov, Isaac, *The Collapsing Universe* (New York: Simon and Schuster, 1977).

Barrow, John, and Silk, Joseph, *The Left Hand of Creation* (New York: Basic Books, 1983).

Bartusiak, Marcia, *Thursday's Universe* (New York: Times Books, 1986).

Calder, Nigel, *The Key to the Universe* (New York: Viking, 1977).

Davies, Paul, *Superforce* (New York: Simon and Schuster, 1984).

Ferris, Timothy, *The Red Limit* (New York: Morrow, 1983).

Gribbin, John, *In Search of the Big Bang* (New York: Bantam, 1986).

Pagels, Heinz, *Perfect Symmetry* (New York: Simon and Schuster, 1985).

Parker, Barry, *Einstein's Dream* (New York: Plenum, 1986).

Parker, Barry, *Search for a Supertheory* (New York: Plenum, 1987).

Silk, Joseph, *The Big Bang* (San Francisco: Freeman, 1980).

Trefil, James, *The Moment of Creation* (New York: Scribner's, 1983).

Wagoner, Robert, and Goldsmith, Donald, *Cosmic Horizons* (Stanford: Stanford University Press, 1982).

Weinberg, Steven, *The First Three Minutes* (New York: Basic Books, 1986).

Index